JESUS FOR FARMERS AND FISHERS

. . . and Farmworkers, Food Service Workers, Ranchers, Lobster Catchers, Sheepherders, Cowboys, Foragers, Orchard Keepers, Pruners, Shearers, Vintners, Brewers, Bakers, Dairy Workers, Cheese Makers, Malters, Manure Spreaders, Mushroom Hunters, Composters, Seeders, Breeders, Irrigators, Clam Diggers, Truckers, Millers, Vets, Market Vendors, Field Gleaners, Soup Kitchen Ladlers, Food Bank Staffers, Landfill Scavengers . . .

Praise for Gary Paul Nabhan

"Nabhan's holistic look [at our food system] extends to his own life, in which daily work and daily spiritual practice provide balance."

—Utne Reader World Visionary Award

"Lyricism . . . infuses [Nabhan's] prose, a rhapsody tempered by hard botanical science."

—San Francisco Chronicle

"Nabhan's painstaking research has not eclipsed an evident natural knack for storytelling."

—Saveur

"Nabhan teaches ecological lessons to nonscientists through an impressive range of disciplines: ancient history, ethnobiology, paleo-nutrition, ecology, history, anthropology, and more."

—Choice Reviews

Praise for *Jesus for Farmers and Fishers*

"A bold, courageous, and utterly original rereading of Jesus's parables that draws us deep into questions of what it means to stand with those struggling with food insecurity, erosion of cultural identity, and spiritual loss. This gifted ethnobotanist, with his eyes wide open, helps us feel anew the pathos and power of Jesus's teachings about food and life and reimagine the world as sacramental."

—Douglas E. Christie, PhD, author
of *The Blue Sapphire of the Mind*

"Every page of this fascinating book imbues the Scriptures with smell and taste, a living landscape and rich cultural tradition. Nabhan's gift for telling true stories and his keen insight into the practices of extractive economies then and now open our eyes to the gospel imperative of food justice."

—Ellen F. Davis, Amos Ragan Kearns Distinguished Professor of Bible and Practical Theology, Duke Divinity School

"Who better to give us a fresh reading of the Jesus story than one of our leading agrarian writers and practitioners? In *Jesus for Farmers and Fishers*, Gary Paul Nabhan's vast scientific and agricultural acumen melds with a deep contemplative wisdom. The result is one of the most insightful readings of the Gospels I've encountered, read through the eyes of the very people Jesus served: fishers, farmers, bakers, gleaners, migrant farmworkers. Here is a book for today's food justice movement, and for anyone who hungers for restoration of our lands and our communities."

—Fred Bahnson, author of *Soil and Sacrament*, and founder of the Food, Health, and Ecological Well-Being Program at the Wake Forest University School of Divinity

"I am hungry for this book. Gary Paul Nabhan calls us to discover the tastes, scents, and textures of food in the Gospels and encounter the people who grew it, caught it, and cooked it. Nabhan's work plunges us into the way of Jesus that turns things upside down and inside out. The powerful are brought low and the lowly raised up. As Nabhan digs into the complexity and depth of injustice in Gospel times, we're shown stories that interweave with those of field hands and food service workers who provide our food—at great cost to themselves."

—Anna Woofenden, author of *This Is God's Table: Finding Church Beyond the Walls*

"Prepare to be illuminated! *Jesus for Farmers and Fishers* brims with insights that can only come when you join in one person the world's leading ethnobotanist, a major food justice advocate, and a member of the Order of Ecumenical Franciscans. Page after page, Gary Paul Nabhan shows how living closely and practically with land, water, and fellow creatures helps readers appreciate Scripture in ways they never have before."

—Norman Wirzba, Gilbert T. Rowe Distinguished Professor of Theology, Duke University, and author of *Food and Faith: A Theology of Eating*

JESUS FOR FARMERS AND FISHERS

Justice for All Those Marginalized by Our Food System

By Gary Paul Nabhan,
known as Franciscan Brother Coyote

BROADLEAF BOOKS

MINNEAPOLIS

JESUS FOR FARMERS AND FISHERS
Justice for All Those Marginalized by Our Food System

Copyright © 2021 Gary Paul Nabhan. Printed by Broadleaf
Books, an imprint of 1517 Media. All rights reserved.
Except for brief quotations in critical articles or reviews,
no part of this book may be reproduced in any manner
without prior written permission from the publisher. Email
copyright@1517.media or write to Permissions, Broadleaf
Books, PO Box 1209, Minneapolis, MN 55440-1209.

Cover image: shtonado/istock
Cover design: James Kegley

Print ISBN: 978-1-5064-6506-7
eBook ISBN: 978-1-5064-6507-4

Printed in Canada

CONTENTS

This book is dedicated to
all farmers, fishers,
and Franciscans.

The kin-dom of heaven

    is like this: a treasure hidden

        within a field of golden grain,

      which some folks accidentally stumbled upon

and then covered up with soil and brush

      to keep safe deep in the ground

          where it had been originally found.

In their joy, the discoverers

    go out and sell everything

they have ever possessed

      so that they may protect

         and prosper in that treasured place.

INTRODUCTION

t is time for you to taste and see. It is time to smell and listen.

Open your taste buds and eyes as well as your nostrils and ears. As you do so, try to imagine Jesus as a child, one who is beginning to explore this world. Picture him as he comes into his first sensory contact with the delicious bounty of the Holy Lands and the Sacred Seas:

The mildly sweet flavor and semifirm, flaky texture of Saint Peter's fish. It is a tilapia special to the Sea of Galilee.

Nutty, whole-grain flatbreads baked on hot stones or in small wood-fired ovens. The dough for breads came from ancient grains such as barley, einkorn, and emmer wheat. After baking on both sides atop hot stones, the flatbreads were eaten while still warm, still carrying the fragrance of olive branches lingering in the wood smoke.

The pale-green figs of the Mediterranean basin—sweet and delicate in texture and yet so capable of enduring the harsh environments of the desert as they rooted themselves in the slightest crevices of barren cliff faces.

The oil-rich olives, sharply bitter in their flavor until salt, water, and time soften and sweeten them.

The pomegranate, with its tough, leathery skin on the outside. Crack it open and you discover its treasure trove of carmine-colored arils hidden inside. Their moisture-rich packets of sweet juice and crunchy kernels are like a dream to the pilgrim's tongue, which is parched and fissured.

When we taste and see, smell and listen, as the young Jesus did, our senses assure us that the earth that the Lord gifted us is a good place in which to dwell.

Our Creator gave his only begotten Son to this precious world. To be incarnate in it. To be with us *here*. Who can doubt that Jesus of Nazareth took pleasure in these flavors, fragrances, textures, and colors? He may have been poor, but he was no puritan who disavowed his senses.

He offered every passerby a place at the table, regardless of their race, faith, social status, or political stance. Together with his nearest neighbors and wayfaring strangers, Jesus celebrated the sensory abundance of the creation with a fervor and elation usually reserved for families joined at a wedding feast. As though Jesus had to remind us what a delicious, sensuous world we enjoy!

But just what kind of place and time was Jesus himself born into?

For starters, the rural landscapes and seascapes of semi-arid Galilee were *peopled* places more than pristine wildernesses. They were *cultured* as much as a bowl of yogurt is cultured, as much as grapes are fermented in large clay pots.

Trammel nets, or fish traps, were set out in waters where a freshwater spring wells up into the salty sea. And patient orchard keepers might fertilize a seemingly barren fig tree with donkey manure just to see if they could coax it into fruiting once more.

To be sure, the people of the rural villages in Galilee were not just about catching fish, baking bread, and fermenting wine. They also had to deal with the trinity of poverty, pain, and oppression. In Jesus's day, job-seeking farmworkers might stand for hours with their sickles or scythes in hand, waiting in the shade of a carob tree on the edge of the village for some foreman to hire them. A crooked manager of hired harvesters might arrive, hoping to cheat both his boss and the workers by undervaluing their work. An aristocrat might have demanded the payment of a docking fee from all the fishers who hauled their catch ashore at "his" landing area. By such political maneuvers, he might succeed in closing down every other dock to further enrich himself, exploiting the rural families who had freely used the harbor for generations. Meanwhile, a local vintner might cut corners after suffering from a poor harvest of grapes by paying off creditors with diluted wine and pouring the mixture into old, fragile wineskins.

To put it simply, some things have not changed much since Jesus's day two thousand years ago. But keep in mind that the struggles in Galilee were precursors to what has since happened in so many other communities of rural African, Asian, Latino, and indigenous farmers and fishers as a result of globalization. Why was Galilee on the front lines at that time? Because its fertile lands, long growing seasons, and bounteous waters made it one of the most productive breadbaskets between the Tigris-Euphrates river system and the delta of the Nile. And yet their products were much more accessible to Rome and Athens by horse cart or ship than those from Egypt or Mesopotamia. It was only during the time of Jesus that Galilee became a flash point for agrarian conflicts that would later reverberate across the Roman Empire.

Galilee was not some bucolic countryside where peasants pursued a romantic agrarian ideal. It was rife with difficult conflicts and gritty problems: upheavals in land-ownership and use, inequities in food access and production, and the usurpation of resources by rich absentee owners and commodity brokers. Galilee was a crossroads where Jews, Romans, Greeks, Nabateans, Phoenicians, Samaritans, and

Syrians lived in tension with one another. Many of those tensions flared when a large and voracious empire surrounded them, insinuating itself into every aspect of their daily lives. The Roman Empire worked to extract every calorie of food, wine, or oil it could squeeze out of the land and sea. It endeavored to grab every coin out of the purses of every peasant there.

So do not be surprised if the tensions in our own times seem to mirror those that Jesus saw his disciples dealing with twenty centuries ago. We now know that the Galileans were suffering from one of the worst farm and fishing fleet crises recorded in the Western world. This dire agrarian crisis devastated the Middle Eastern peasantry, affecting the people around the Sea of Galilee at the very time that the feet of Jesus first touched its shores.

Why did the farmers and fishers of that era feel like an unprecedented crisis was pulling them asunder? In *Jesus and the Peasants*, New Testament scholar Douglas Oakman notes that their destitution had become crippling due to the increased burden of debt they were shouldering under Roman rule. This included new taxes, tributes, and land rents, as well as tithes and religious dues. Many peasants had been evicted from their lands and were forced into beggary, prostitution, or tax collection—some of the few occupations not tied to the land.

As often happens in history, that kind of economic instability led to political unrest as well when the peasants had too much of this oppression. Each time there was a change in government, they tried to voice their concerns and leverage some change, but the political power mongers would crush them immediately. In *Jesus: A Revolutionary Biography*, John Dominic Crossan says that the death of Herod the Great in 4 BCE, right around the time Jesus was born, sparked rebellions across his territories. The Roman government implemented harsh measures to control the population by force. Crossan says that if the peasants were lucky, "they lived at subsistence level, barely able to support family, animals, and social obligations and still have enough for the next year's seed supply. If they were not lucky, drought, debt, disease, or death forced them off their land and into sharecropping, tenant farming, or worse."

Jesus intentionally chose to carry out his mission among this most motley of crews: homeless scavengers, migrant farmworkers, itinerant gleaners, and those who appeared out of nowhere to help gut the fish that came in as the catch of the day. As soon as he arrived in the communities of Galilee, he seemed to grasp the irony that the very people who brought their urban neighbors their bread, wine, and fish had joined the ranks of the marginalized. Within the span of a single generation, the Galileans had seen nearly two-thirds of their annual catch at sea and their harvest from the land shunted from them in order to support the urban elite and a distant aristocracy.

Jesus could not and did not simply offer his followers sugarcoated condolences to soothe and to placate them. Nor did he use the pie-in-the-sky revolutionary rhetoric of zealots who could rile them up but then do little more than lead them to slaughter by the military.

Rather than merely raging at the machine, Jesus offered game-changing principles that he sensed could avert further conflict and help people regain their dignity. He focused more upon strengthening a sense of justice, dignity, hope, and resilience through stories that continue to have staying power centuries after they were first told.

That is why I am writing this book. I am convinced that the agrarian parables, proverbs, and aphorisms of Jesus have an uncanny relevance for us today. Just as there was clearly a crisis for farmers and fishers in Jesus's time, many in our own society are now slipping into crises of similar magnitude, ones that have begun to sink our fishing fleets in debt and knock the legs out from under our on-farm workforce.

I, for one, can no longer deny or stay distracted from the widespread suffering I am seeing and hearing all around me. The underlying power dynamics of these economic disruptions and racial injustices today hauntingly echo those that undermined agrarian communities all across the eastern Mediterranean region during the era when Jesus walked upon this earth and waded in its waters. It was the very moment in history when he spoke to such concerns in a canon of parables and proverbs that we ironically call the "good news"—the gospels—for they may offer us a way out of such suffering and sadness.

That is why it is worth listening anew to the cautions and wisdom Jesus offered farmers and fishers of his own era. Undoubtedly, his parables became his most memorable commentaries meant to lift up the indebted and enslaved.

Is it not miraculous that these parables are still relevant two thousand years after Jesus first let them loose on our lonely planet? At the very least, such stories bear witness to the lives of those who depended daily on the harvesting and marketing of fresh produce; the managing of weeds and pests; the caring for alkaline soils and scarce water; the darning of torn nets; and the dodging of torments, tariffs, and fists by tough-minded day laborers.

There can be little doubt that Jesus felt a special affinity with those who produced, processed, and put up food. In the contemporary Arabic of the Levant (including Galilee), these food-producing members of the "peasantry" are still called fellaheen collectively, as they may have been during the era of Jesus.

You can almost hear Jesus shouting out to them, "Hey, fellaheen, come join us!" welcoming them and feeding them with the best he could muster.

From the all-too-brief accounts of his life that survive, we know that Jesus joined them in their fields and in their boats. In the quiet and coolness of their homes, harbors, and orchards, he spoke directly to their fears and humiliations. Whenever he spoke to them, he always did so with intense compassion, as if he held their best interests in his mind and his heart.

Of course, the fate of farmers and fishers was not his sole concern. He cared as much for children, widows, cripples, lepers, prostitutes, the blind, and the lame. His own family members were "carpenters," of a social class referred to as tektōn—a catchall term from ancient Greek for artisans who built things with their hands, whether they be stonemasons, woodworkers, cartwrights, thatchers, plasterers, or blacksmiths. Considering that Jesus came from a family of artisans who were neither farmers nor fishers, the content of his parables is striking. Nowhere in Jesus's parables do you find images gathered from his own labor in the trades of building homes, walls, tables, granaries, or wells. His stories do not cover a single struggle embedded in his or his father's profession but focus instead on those at work on the land or in the seas.

Could it be that Jesus's heart reached out to farmers and fishers in a dramatic way, not because he romanticized them as the "salt of the earth" but for another, more troubling reason?

Was it because they had recently joined the ranks of the most oppressed and dispossessed in his society?

There's another thing missing from Jesus's parables that you might expect to be there, given his location and time. While Jesus lived his last few years completely immersed in a world where peasant farmers and freshwater fishers were among his friends and neighbors, he was also within easy reach of the increasingly urbanized world of Romans, Greeks, Nabateans, and Phoenicians. The city of Sepphoris, one of the hubs of commerce closest to the Sea of Galilee, was within walking distance from where he grew up in Nazareth. Within the span of Jesus's brief life on earth, Sepphoris became a burgeoning metropolis with perhaps as many as eight to ten thousand residents and itinerant visitors.

Sepphoris sprawled out from its temples, theaters, and colonnades until its city limits were less than four miles away from the home of Joseph and Mary. Those who lived in its palaces demanded massive amounts of food and drink from the surrounding countryside to fuel their insatiable appetites. The extractive power of this commercial hub forced farmers and fishers to become "miners" of all the wealth that could be pulled from the soil and the waters. The city was renamed Autocratoris by Herod Antipas, and most of its residents came to make their claims in an imperialistic economy predicated on extraction and taxation. (Don't you just love the audacity of autocrats naming a city to honor their dubious profession?)

To the frustration of many historians, Jesus never mentioned this metropolis in any of his surviving commentaries. While he does make mention of several professions performed in this city, their influence is alluded to only indirectly. His parables are about the farmers and fishers who found their ways of life upended in the first century, not about those who profited by that upending.

When he preached, Jesus chose to express how deeply troubled he felt because of the crisis that had rather suddenly overwhelmed the region's food producers. But he could not do so openly, challenging the government or its unjust economic system. No one under the thumb of the Roman aristocracy and military could speak directly about the political dominance, economic enslavement, and spiritual malaise they were experiencing. Anyone who "protested too much" could be captured, chained, whipped, or even crucified. To prevent this from happening to him, Jesus could only discuss certain politically or economically charged dimensions of this crisis through veiled allusions, or parables.

Fortunately for us and for the peasants of his era, Jesus was a master craftsman of a metaphorical language. The commentaries that he offered in light of this agrarian crisis were verbal riddles or puzzles that peasants had to ponder in order to solve. They were not recipes for quick fixes. Jesus did not stoop to simplistic political prescriptions.

At first glance, many of his stories ostensibly focused on some of the most mundane topics regarding farming and fishing: the weeding and sickling of grain crops, the pruning and fertilizing of fruit trees, and the mending and placement of fishing nets. For the most part, he evaded accusations by the Jewish power elite that he was "sowing wild oats" of dissent or "rocking the boat" of hierarchical relationships.

Jesus himself went to reach out to the fellaheen on their own ground while they labored on land or at sea—to offer them a fresh way of looking at their own lives. He engaged people where they were, in the field or the harbor, and he spoke to them in humble, simple terms. Although he was the Son of God and had the full authority to speak as any rabbi spoke, he preferred the gritty language of farmers, farmworkers, and fishers.

Franciscan teacher Richard Rohr has noted that Jesus had a poet's peculiar capacity to take something mundane and then turn it "upside down" or "inside out" to reveal a transcendent truth. The earthy language Jesus wove into his sermons was so lovely and unforgettable that it has kindled a warm glow in human hearts for more than two thousand years. From these metaphors of dead branches pruned from fig trees, of wheat grains winnowed from the chaff, we learn fundamental truths about God.

Let's now listen to those timeless stories that still speak to us today, for they offer a vision of how we might deal with agrarian crises of our own time that are devastating farming and fishing communities all around the world. Let them remind us that the time-tested parables of that lonesome contemplative can still lead today's farmers and fishers of all races, genders, and cultures toward a more exuberant life and abundant harvest.

PROBLEMS
AND PARABLES

Jesus knew that the people who followed him and listened to his preaching were a perceptive and intelligent lot despite their lack of formal education, and he spoke to them in a way that never sounded esoteric or "otherworldly." Through his parables, he celebrated their earthly knowledge, drawing profound truths from everyday things they knew: a labor dispute, a lost coin, a crop that failed to bear fruit.

His stories are just as relevant today despite the two thousand years and deep cultural divides that separate us from Jesus. That's because many of the problems ordinary farmers and workers in his day faced are the same problems happening in our own time.

Let his stories beckon us to walk in their sandals—not our own fancy shoes—to imagine what it must have sounded like and felt like when Jesus burst into memorable stories.

1

WHAT WAS HAPPENING
THEN IS HAPPENING *NOW*

Are you beginning to sense that what was happening then, in Galilee, is also happening again right now, in the heart of America?

Have you walked through fields in fall where more thistles and bristly seeded weeds are growing than harvestable crops? Have you gone out to fish where the only things left to catch are small fry?

Can you imagine a time when farmers and fishers of marginalized races, cultures, and creeds realized that the resources they required to make a living were in steep decline? Can you feel how this dilemma was propelling them into fiscal bankruptcy, political disempowerment, emotional despair, and spiritual disillusionment?

When it comes to farmers and fishers in crisis, every old struggle inevitably comes around once more even if they wear different boots or sit on a new tractor seat. History

Problems and Parables

repeats itself again and again. But Jesus, through it all, offers fresh hope for us to break the cycle.

In Jesus's time, many peasants who once farmed or fished for a living had thrown in the towel. They had left their villages to take menial jobs in the cities, reluctantly and sullenly catering to the needs of the elite.

In fact, as cities were burgeoning, developers, money-lenders, mortgage bankers, and politicians took some of the best arable lands that farmers had left. There, they established plantations exclusively for olives or spices sent off as export crops to Rome. On the shores of the Sea of Galilee, they constructed fancy villas where fishing processing had formerly occurred around the best natural harbors. The Romans depleted the sea of its abundance by overfishing and contamination.

The new urban elite did not emerge from local families but from those in Jerusalem who had forged power alliances with foreigners. The Romans placed these well-connected Jews and Nabateans as intermediaries in key political and economic positions within all communities. The expansion of urban markets intensified food production to the degree that it depleted the soil of its former fertility.

Those fecund soils had been nurtured by the Galileans' ancestors over centuries of tireless, careful stewardship. These trends alarmed the best of the farmers and fishers who struggled to remain in their places. Still, many of their neighbors opted to be bought out, simply shoving money into their pockets so they could seek better pastures and harbors elsewhere.

But where could the displaced farmers or fishers go to escape such pressures? The economic empire of the Greco-Roman elite was all around the Middle East, and it took advantage of rural families everywhere.

Tax collectors demanded payments for the use of any fields or boat docks that peasants had inherited. Brokers hit them again for a percentage of the food they took into the city to market. Toll collectors exacted a price for any produce or seafood they moved into the city on privatized roads.

Whenever farmers or fishers suffered poor harvests, moneylenders loaned them money at high interest rates to get them through to the next year, further increasing their debts. While rabbis officially condemned such punitive usury, it was practiced in every corner of Galilee.

As the debt levels became too high, the moneylenders assumed the equity of peasants' farmlands or fishing boats unless they paid off both the principal and the accumulated interest. The farmers and fishers then became sharecroppers, journeymen, or contract workers for the elite, some of whom treated them like indentured servants.

Can you sense why so many Galileans grew despondent? They had lost control of the small properties and the larger "commons" that had nourished their clans for many generations. Men in particular grew sullen, drinking more wine or taking narcotics that kept them in a stupor and dulled their physical and emotional pain, at least for the moment.

When family members expressed frustration that their husbands or fathers were acting as if they were humiliated or emasculated, the depressed men lashed out against them, abusing their wives or children. Upon realizing what they had done, some of the men were overcome with grief and guilt. A few opted for suicide.

Of course, I am recounting a historic moment of agrarian crisis that occurred roughly two thousand years ago. It was not the first agrarian crisis on the face of the earth, nor was it the last. A hero of mine, theologian Ellen F. Davis, has perceptively described other such crises in her remarkable book *Scripture, Culture, and Agriculture*. According to Davis, the terrible crisis that faced peasants in Galilee in Jesus's time was reminiscent of another, perhaps milder one that had started nine centuries before and continued for many decades. The monarchy and the aristocracy had appropriated family farms and undermined the traditional system of land inheritance.

The prophet Isaiah, speaking of this time, denounced the greedy elite as "the ones who, usurping homestead after homestead, amass field to field until there is no more room for others so that they dwell all by themselves in the midst of the land" (Isaiah 5:8).

If these trends sound familiar, well, they did to Ellen Davis as well.

Davis recognized common themes among the agrarian crisis of today, the one that flared up nine hundred years before Jesus, and the one during Jesus's lifetime that transformed farming and fishing towns around the Sea of Galilee. In many of these farming and fishing communities throughout time, within the current struggle of global communities, to those in Jesus's time oppressed by Roman systems, to the earlier plight of agrarian communities in the time of the Maccabees, we see the same power structures, the same disadvantaging of farmworkers. After seeing what was happening to his neighbors who labored as farmers and fishers, Jesus could no longer remain silent. As he found his voice, this humble Jew began to make waves that still ripple toward us today.

As soon as Jesus landed in Galilee, he could sense how the urban elite disparaged and dismissed the fellaheen as hayseeds, as naive nobodies who could easily be duped. These peasants were treated as a gullible people, a culture apart from the more sophisticated Jews in the Old City of Jerusalem.

Worse yet, the Roman capitalists and imperial troops regarded them as expendable minions. It had become increasingly difficult for peasants to express their outrage over what had occurred, since the lands and seas of Galilee were occupied by Roman troops.

Galilee had essentially become a zone of sacrifice, a source of raw resources that fueled an extractive economy controlled by the "resource sink" of the Roman Empire. Most wealth and resources that passed through small, regional outposts of commerce were destined for Rome.

For example, pickled olives and fermented fish paste from Galilee were suddenly being hauled by oxcarts and ships all the way to Rome itself. The military occupation of Galilee supposedly safeguarded residents there from outside marauders, but it spent more of its force on exploiting resources, breaking up protests, and quelling other civil unrest, just as all police states do.

It wasn't just in Isaiah's time, or in Jesus's time, that Davis glimpsed patterns of oppression. While she was delving into those agrarian crises from the Bible, she read Wendell Berry's classic *The Unsettling of America* and recognized the same patterns happening in our time. Berry wrote eloquently about the roots of the US farm crisis of the 1980s.

What was happening then is occurring again now: this profile of a food-producing landscape in the throes of agrarian crisis is also apparent today. It can be seen in hundreds of rural counties across the North American continent and in dozens of other countries around the world.

Farmer Victor Davis Hanson reminds us of this in his fine 1996 book *Field without Dreams*, which chronicles the demise of the family farm. By any yardstick, he says, we are now in "the penultimate state of the death of agrarianism, the idea that farmland of roughly like size and nature should be worked by individual families." In agrarianism's place is a concentration of farm wealth in the hands of a tiny minority. Less than 1 percent of farmers now account for the majority of farm income, while 88 percent of agrarian producers earn less than $20,000 a year directly from the food they harvest.

Since that trend still persists more than two decades after Hanson revealed it, why are we even surprised that the rates of suicide, alcoholism, and opioid addiction among

American farmers are at an all-time high? The same trends hold for other agrarians scattered all around the world. In the US, farmers suffer some of these maladies at rates twice the national average for adults. *Newsweek* welcomed us to a new era of "death on the farm" in a 2014 issue: "By the 1980s . . . farmers started defaulting on their loans, and by 1985, 250 farms closed every hour. That economic undertow sucked down farms and the people who put their lives into them. Male farmers became four times more likely to kill themselves than male non-farmers, reports showed. 'In the West, the guys were jumping off silos,' says Leonard Freeborn, a horse farmer and agricultural consultant. Since that crisis, the suicide rate for male farmers has remained high."

How did we get ourselves into a predicament so pervasive that it now afflicts at least a third of the world's human population? Today, small farmers, farmworkers, fishers, herders, and food service workers are among those most likely to suffer from poverty, hunger, and food insecurity. Those workers who happen to have black, brown, or olive skin fare far worse.

To this day, four-fifths of the world's food is still produced by family farmers, not large corporations. That means that four out of every five meals served in homes and communities around the planet still come from small, family-owned farms and not from the corporate-controlled feedlots, slaughterhouses, and greenhouses that employ thousands of poorly paid contract workers.

By a small farm, I mean a piece of property now minuscule by Midwestern Corn Belt standards. Even today, seven out of ten farms worldwide are less than one hectare in size—which is probably about what they were in Galilee, Palestine, and southern Lebanon when Jesus sauntered through the countryside there.

That is the scale of farms most vulnerable to globalized commodity markets, taxes, tariffs, and savings and loan frauds. While these small shareholders still grow most of the world's food today, they fear they are now facing an agrarian crisis of "biblical proportions." In the US, the ratio of farm income over expenses has been flat since 1950, while living expenses and health-care costs have risen dramatically. And while the cost of nearly everything we put into our mouths has risen tenfold since World War II, the price a farmer can sell a bushel of corn or wheat for has nearly halved since then.

This is more or less the same fix that the farmers and fishers of Galilee found themselves in two millennia ago. No wonder they were curious about the commentaries by the enigmatic prophet who had appeared in their midst. At last, someone had the guts to address the world of hurt that they found themselves in.

So how did Jesus respond to this situation? What message did he have for the farmers and fishers who were displaced and suffering during his time—and what message can we, by extension, receive and apply to our own challenges today? Let's take a look at three notable aspects of Jesus's approach.

First, it is clear that he did not side with the elites. He did not spend his time courting the greedy and the powerful. Jesus chose to work, hike, eat, and drink not only with farmers and fishers but also with indentured servants and foreigners, tax collectors and scribes, and so-called harlots and prostitutes, many of whom had been slaves of sex traffickers before they broke free.

Second, he did not embrace violence. Jesus may have allowed some of the zealots of the protest movements of his era to join his ranks, but he himself was not a card-carrying participant in any of their overtly confrontational political strategies.

In *Rabbi Jesus*, Bruce Chilton reminds us that Jesus and his disciples did not take up arms or affiliate with the political movements of the time. He valued peace. When Jesus was arrested, a badly shaken Simon—now called Simon Peter or simply Peter—pulled out a knife and cut off the ear of a soldier while trying to protect his teacher from torture, imprisonment, and death. Jesus did not condone this action. In fact, it disturbed Jesus so much that he immediately focused not on his own pain but on that of the soldier, turning his efforts toward healing the man's ear.

That's it—all the violence we know that was associated with Jesus and his first disciples. One knife.

No sword-toting, gun-toting, ax-toting, bomb-toting militants emerged out of the motley crew of peasants who first gathered around Jesus in Galilee. Nevertheless, Jesus had far more influence on the destiny of Galilee and the world at large than most zealous militants could ever imagine.

Third, he did not revile his opponents or those who persecuted him. Paradoxically, he changed the political debate of his era while transcending politics itself, and did so without ever abandoning his defense of the poor, women, the elderly, the abused, and the disabled against the structural injustices of their era.

Jesus fully embodied Eckhart Tolle's maxim that "you cannot solve the evils of the world on the same level of consciousness that created them." Even so, many activists working for social justice, food equity, environmental improvement, and fairer immigration today still try to confront the evil in the world by vilifying anyone who disagrees with them, just as most feudal warriors have done for centuries. If we are to see social and agrarian justice truly advance, we need to consider how to act not only within but beyond protest movements, as Jesus did.

So those are three things Jesus did *not* do as he addressed the issue. What he *did* do was preach the gospel in parables, using everyday examples that farmers and fishers would understand.

To understand this approach, we have to ask not only "What did Jesus mean?" but also "How did his peasant neighbors hear him?" In this respect, we might wish to employ an agrarian lens to understand how Jesus's listeners heard what he said about their historic farming and fishing crisis, in ways that offer insights into our current one.

This agrarian lens for reading the gospels takes into account the traditional ecological knowledge of the farmers and fishers of that era, to whom Jesus was speaking. It also recognizes the political ecology of the first century and the pressures it placed on all who worked in the food economy of the Holy Land.

An agrarian lens allows us to do an earthy reading of most of the parables so that we focus more on the interactions between and among cultures, classes, creatures, and their surroundings. Such a lens into the teachings of Jesus can only enrich us if we begin to fathom the rural social dynamics, the agro-ecological context, as well as the richness of culinary symbolism and the sacramental dimensions of farming, fishing, and food in his place and time.

When we look afresh at the parables through the eyes of Middle Eastern farmers, fishers, herders, and orchard keepers, as we will do in the chapters that follow, we can clearly see that Jesus was offering them both the intangible gift of hope and tangible options for survival. Jesus guided his hearers into rethinking for themselves how to survive and build community at the very moment that they felt overwhelmed by unprecedented pressures. And it seems that those who most deeply "heard" and heeded his words were the most marginalized in his own time.

Such a vision might help us gain greater empathy for those who have struggled to retain their rural livelihoods during every period in history when such agrarian crises have occurred, including our own.

In coming to understand Jesus's parables, you might gain insights through this agrarian lens that will help you find ways to ease the plight of farmers and fishers in our present time. You might become able to better support them in their struggles to survive and fully contribute to our society and its expressions of faith.

So it's time to roll up your pant legs to wade out in the Sea of Galilee. Let's see if you can spot the stories that can be harvested from its waters.

2

THE SOWER

D o you fondly remember stories told to you when you were a child, ones that made you dream or see the world in fresh and wondrous ways? Although children do not necessarily distinguish parables from other kinds of stories, they often discover delight and a sense of new possibilities looming before them. For many adults, the retelling of parables also triggers a kind of healing, for they help us move past our former wounds into greater fullness and hope.

Hope? Well, hope might best be most welcomed when sown in places where the recent harvests have been meager, where hardship has limited the pursuit of happiness. Most adults I know have suffered moments of meager harvests—heartbreaks or humiliation in which they long for health and abundance once more.

So how did Jesus select and shape his parables so that they gave luckless fishers and failed farmers such a visceral sense of hope? How did that hope spread among so many who had already lost their most fertile lands and/or finest harbors to tax collectors or real estate speculators?

The few Galilean farmers who remained on their lands had only their least arable plots left to use for grubbing out a living. The fishers barred from the best anchorages rowed back and forth across overexploited waters.

Nevertheless, Jesus somehow found a way to reach them through his knack for healing their brokenness and inspiring hopefulness.

Jesus could demonstrate such deep empathy for the fellaheen dilemma through his parables even when some of his more elite and urbane disciples could not get the drift of where some of his stories were heading.

Of course, we too may find it hard to fully understand those narratives, for we live in a different time, place, and economy. Details of the routines of Galilean fishing or farming life are lost upon many of our own generations—not for lack of intellectual interest, but for the infrequency of our exposure to such rural livelihoods. Four in five Americans now live in urban settings compared to barely half a century ago. We inhabit a material world quite different than the one in which Jesus dwelled, when 90 percent of all humans were farmers, fishers, foragers, hunters, or herders.

Nevertheless, the challenges faced by farmers and fishers should remain of concern to us for one simple reason:

we eat by the sweat of their brows. For that reason alone, deciphering the hidden meaning of the parables of Jesus can be worth our while.

Fortunately, it is not too hard to get into the swing of what the parables mean, even though they were probably first spoken in the now-endangered language of Aramaic. The imagery and cadence we find in the aphorisms and parables of Jesus are those of a gifted storyteller who reached his listeners through colorful but cryptic symbols, curious riddles, and circular plots that engaged listeners *as participants* in the process of making the story whole.

There was no need for Jesus to stand behind a podium or pulpit to pontificate. Instead, he interacted with his listeners' hearts and minds in a manner that became integral to the story itself.

The only way the story could be made whole and would make wounded listeners whole was by engaging them in deep participation. Whether we regard Jesus as a charismatic street preacher, outrageous performance artist, or erudite rabbi, his storytelling skills had to enchant as much as edify the farmers and fishers, streetwalkers and sinners, of Galilee and Judea.

Participation in these stories is something that we too must engage in if we are to truly "get the drift" of where they might take us.

Try to *actively* listen to a fresh telling of one story that three gospels offered in slightly different versions. See how you might use such stories as springboards to imagine how those who listened to Jesus in the fields and fishing boats of Galilee became empowered to *live out* such narratives:

Hey! Listen up, those of you who think you have ears!

Some craned their necks to see just who it was that was trying to rile them up in the heat of the midday sun. Others yawned, and some winked at one another, for they had heard other street preachers before. Nevertheless, they all leaned in closer so that they could better listen to what this stranger had to say:

A farmer went out to sow,
and from his hand he would throw . . .

He gestured with his hand, as if flinging seeds out toward them in every which way.

. . . a broadcasting of the seeds,
　　　　but most of them landed
far from the sower and too close to the barren road,
and so the birds had little trouble
　　　　Swooping down to devour it.

He crossed his thumbs in front of him then fluttered the fingers on both his flared hands, moving them up higher and higher into the sky, as if blackbirds (or bankers) were carrying away their recently purchased seed stock. The crowd cackled at his antics.

　　　　Some of the seed they cast out
fell where bedrock reached the surface.

He knelt upon the stony ground before them, knocking his knuckles against the hardened earth to demonstrate its impermeability. They heard a low thud. They knew all too well that seeds cannot penetrate very far into compacted earth.

　　　　Even though a scatter of these seeds
　　　　quickly sprouted atop the hard-packed soil,
　　　　their roots could find no moisture in the earth,
　　　　and when the sun desiccated them, they withered and died.

He held his hand up as if trying to block the sun, but still he squinted and sweated before them, unable to shield himself from its desiccating radiation. The sun's unrelenting rays scalded the seeds as they lay on the parched earth.

Others of the seeds he sowed
landed among some thorny brush
that shaded out their emerging sprouts
while their roots were choked out and rendered lifeless.

He grabbed a branch of spiny, tangled crucifixion thorn and forced his fist up through its barbs until the skin on his hand dripped with blood. The people themselves had felt their own arms and legs scratched and bloodied by the piercing of these thorns whenever their crew bosses had forced them to walk through dense thickets of brush along dry desert watercourses.

At last, the sower came to a place
where the earth felt welcoming, full of tilth,
where he could gently fling some seeds into sweet spots
where they made their way to deeper, richer soil.

He knelt down again and used his bloody hand as a trowel, but this time, he brought up fragrant, richly textured, glistening humus from beneath the stones on the surface. He raised it up, then he bowed to the fellaheen who had gathered to hear him. He stretched his other arm out toward them and opened his hand in deference, as if to remind them that they themselves were essential elements for sustaining the fecundity and generative energy of this earth.

> There, the seedlings grew and bore so many fruit
> that well-placed seeds returned
> at least thirtyfold for every one he had sown
> and yields of as much as sixty or a hundred to one.

Jesus stood up, brushed the dust off his hands and his knees, and began to walk away from the crowd. But as his audience got up to stretch and go, he suddenly turned around and looked at each of them with his piercing eyes. As he shifted once again to walk away, he cupped his hands up to the sides of his head and wiggled them, laughing and shouting.

> Now, for those of you who have ears, let them hear!

Have you ever noticed that the parable of the sower (paraphrased previously) has a curious feature to it, one that few theologians have addressed? It is about the variability of soils in a single field—the substrates that nourish not only the lives of crop plants but our lives as well. Although farmers are concerned with this matter all the time, hardly any scholars have recognized its importance.

The sower, even when he had limited seed supplies due to the rising costs of farming during his era, did not restrict his "broadcasting" of seeds to the most fertile soil available. Instead, he sowed his seed stock on stony ground, on barren surfaces close to a road where birds might get it, or amid thorny weeds. He also tossed seeds into the deeper, moister soil where its tilth, accumulated over the ages, could offer favorable conditions for survival and growth. It is likely that these patches of fertile soil compose just a small portion of the land left available to the sower. But beggars (and impoverished sowers) cannot be choosers. In short, economic and environmental constraints may have forced the sower to plant some of his seeds on whatever marginal land was available, not just on the best.

The ancient landscapes of Galilee included true deserts as well as milder, semiarid Mediterranean habitats. Yet both suffered from highly varying rainfall regimes. The drier an agricultural landscape is, the more variable its productivity is from year to year and from site to site. It's a tremendously insecure environment. It is not an exaggeration to claim

that the Galilean grain farmers had to master the art of dealing with uncertainty of all kinds—adapting to the ups and
downs in precipitation, sudden freezes and lingering heat
waves, heterogeneous soils, and unpredictable outbreaks of
crop pests like locusts and diseases like wheat rust. Not to
mention changing political regimes.

They had to learn what theologian Alan Watts once
called the wisdom of insecurity. Today, some agronomists call
it resilience thinking.

So why might Galilean farmers sow some of their precious seed in less-than-satisfactory places? Was it that they
sensed they should not put all their eggs into one basket?
Was this their bet-hedging strategy: to mix it up in order
to keep it responsive?

The sower did not have much of the best ground where
tilth and fertility generated high yields in the best of years.
He knew that hard times would come again and that he
should be prepared. He did so by selecting a diverse array
of seeds that could endure all sorts of uncertainty. Some of
them could make it through even on the most meager soil,
even under the thorniest shrubs, even during the worst
drought conditions without any pampering. The sower
could have been selecting seeds for their adaptability to
ever-varying conditions as a strategy against uncertainty.
In other words, he sought resilience and greater stability

rather than simply aiming for high yields that might only come in the best of years.

In fact, preparing for uncertainty is what farmers of einkorn, emmer wheat, and barley have mastered in the Middle East over thousands of years. Instead of growing a single "superior" variety in pure seed lots, traditional farmers have tended to sow "multiline mixtures"—amalgams of varieties with no attempt to breed or select them for uniformity and purity.

In the surviving farmscapes of the Syrian deserts not far from Galilee, barley researchers have recently confirmed the lasting relevance of what peasant farmers there have done all along. A heterogeneous mix of seeds provides for a more stable yield from year to year than any pure line of a single "elite selection" of seed could ever do.

It's not about purity—a conceit that Jesus was wary of—but about *diversity and adaptability*.

When the field researchers tracked the yields of dozens of traditional Syrian barley varieties that farmers often mixed in their bags of seed, they discovered that this heterogeneity has staying power. That is because a few of the traditional varieties in a typical mix tolerate cold better than others, while the majority have a good tolerance to heat. Some mature quickly, while others develop their grain more slowly. Some barley stalks tend to "lodge" or fall over whenever there are strong winds, while others are wind-resistant even on shallow, stony soils. Certain seedheads of Middle Eastern barleys are full of barbs to deter birds, while others exhibit disease resistance that most barleys lack.

Do you get the picture? The diversity of their "motley crew" of barley seeds serves Middle Eastern farmers well in the face of exacerbated aridity and heightened uncertainty. For broadcasting seeds like grains, you want to have a mixture in your bag, not a uniform seed stock.

When the field researchers compared harvests from a pure line of improved seeds with a multiline mixture of heritage barley seeds grown under stressful conditions, the mixture averaged a yield nearly two-thirds higher than the pure line over a series of years. The mixture also achieved the highest ultimate yield of any barley crop grown under stressful desert conditions. While the pure line yielded no grain at all in the driest year, the mixture's worst yield was about the same as the average of the pure line when grown under stress! Curiously, the mixture of traditional barley seeds also rendered breads and soups that Syrian families found to be both beautiful and delicious because their grains were of many colors, textures, and flavors.

There are parallels between what a traditional farmer of resilient desert crops chooses to sow and what Jesus communicates here about the wisdom of insecurity. Over and over again in his teachings, Jesus dismissed or sometimes defied any overly restrictive "purity" regulations and taboos of the priestly class.

Here Jesus invites us to relax our concerns for purity and become more receptive of adaptability and diversity in the face of uncertainty. He wreaked havoc on the Hebraic concepts of purity and food taboos on other occasions recorded in the good news. On one occasion when Jesus was challenged about whether he was breaking boundaries with regard to what he ate, he simply replied that the problem was not what goes into a person but what comes out of him or her. Jesus opted to eat and drink whatever was available on whatever day (including the Sabbath) but always did so after washing his hands and offering prayers of gratitude. Rather than dining with bluebloods and elected officials, he was more prone to sup with refugees, ragamuffins, rounders, vagabonds, and wastrels. He was nourished by potlucks more than by banquets.

Jesus did not recruit his disciples by carefully selecting them through job interviews at an employment agency. He found them in the bottoms of boats smelling of fish slime and in the shadows on the rough edge of town. And when Biblical historians count all his so-called disciples mentioned in the gospels, there aren't just twelve. His followers included *dozens* of dirt farmers, scrappy fishers, edgy outcasts, clandestine zealots, recovering tax collectors, abandoned women, and hapless harvesters with short-handled hoes.

Jesus took all of them in, even the one who later betrayed him. He did not simply sow his words into the "seedbed of the elite" where highly educated scribes and rabbis might receive them. Instead, he seeded his messages on the margins of society, in some of the poorest social substrates in all of Galilee. That may be because his rude bunch offered him a heterogeneity of views that Jesus could not have heard in the rarified society found in the temples and courts of Autocratoris. Like the sower in the story, who was rewarded for sowing his seeds in a variety of places rather than holding all of them back for exclusive placement on prime agricultural lands in Galilee, Jesus opted for heterogeneity. He simply refused to "buy in" to the supposition of superiority or exceptionalism of the elite in society, the purity of seeds, or the exclusive use of optimal lands.

It is probably not good for you and me to be too purist or too picky. It may stand in the way of our receiving the grace of God's abundance here on earth!

FISH STORIES

Count them. Narratives about fishers and the fish of the Sea of Galilee appear in seventeen passages within the gospels.

The call to the four fishermen, the calming of the storm, the loaves and fishes, Jesus walking on the sea, the feeding of the four thousand, and the resurrected Jesus asking to eat broiled fish are just a few of the stories where water-born creatures and salty characters of the seas come alive.

In addition, there are seven gospel passages with aphorisms that use fish metaphorically and eight dialogues in which fish come up for air!

Let's dive into just a few of these rich stories.

3

SEVEN SPRINGS

Try to imagine a scene from centuries ago when someone hooked a mystery fish. By now, many of the details of that catch are forgotten except among today's fishers in Galilee: they remember what kind of fish it was, the kind of fishing net used, and where the fishing hole was located. We know it was not far from Capernaum on the shores of the Sea of Galilee. Perhaps the name or exact location of this site was never divulged, for fear of prompting overfishing by undisciplined, exceedingly desperate, or unreasonably greedy fisherfolk. No size or weight of said fish was ever recorded. No trophy specimen was ever mounted on a wall.

But when this fishing story was first told, it must have been richly significant and fully fleshed out, for its filleted fragments have survived in three of the four gospels. We don't have all the details, just the barest of bones—the skeletal remains of wriggly stories that once had muscle, cartilage, fat, skin, and scales surrounding them. The disciples of Jesus who lived in Galilee probably took the knowledge of certain facts for granted. They were unnecessary to record in any detail because every Jewish peasant in the region was fully aware of such particulars.

That leaves us to do the work of salvage archaeologists, attempting to use other "external" cues to put some flesh back on the bones of the stories of Jesus's life, gestures, and words. So we draw on knowledge of physical geography, ancient place names, folklore, ecology, and phenology—the study of the life cycles of animals and plants—to reconstruct what was going on in Galilee two thousand years ago. Let's look at Jesus's first fish story with those lenses, using our imaginations to envision what a fleshed-out account may have sounded like to those early Christians of the Levant who listened to the stories as Jesus *performed* them. If we *remythologize* Jesus's stories by giving them a richer sense of the place and time in which they were told, we can dive into deeper waters.

Some say it happened the first winter after Jesus came to the freshwater Sea of Galilee, at a spot near Capernaum. Imagine that just after sunrise early one morning some fishers recognized Jesus walking on the shoreline not far from their usual harbor. Some then sought him out, hoping to get the latest word about the fate of John the Baptist—currently locked up in jail—and his prophecies about God.

A few of them had once seen Jesus in animated dialogue with the Baptist down by the River Jordan, but now he appeared crestfallen, standing alone on the shore where people regularly congregated to purchase fresh fish. He was not far from two boats of luckless fishermen who had come into the moor to clean their nets and rest after a full night out on the water without a single catch.

The exhausted fishermen waded in the shallows to scour their linen nets clean of algae and sticks so that they would dry out instead of rot. As they did so, Jesus slipped into the boat belonging to Simon and his brother, Andrew, sons of Jonah. He whispered something to Simon that no one else could hear.

They could see by his outfit he was no kind of fisher, but he also didn't seem like a hijacker of boats. So Simon let him sit there in the boat and even pushed him out a ways into the placid waters of the lake to get away from the crowd. The boat was hitched to shore by a long rope of hemp anyway, so little could happen while Simon and the others finished cleaning up their linen nets.

Amused by the sight of a dry-lander sitting in a boat, and intrigued by the rumors that this stranger had been a beloved colleague of John the Baptist, the crowd swelled.

The Baptist's heir apparent could no longer travel incognito in these parts. More people gathered along the shore to see what he was up to.

Jesus seemed unperturbed at first, simply sitting quietly in the boat as if in a trance. But when he looked up and saw hundreds of eyes on him, words began to gush out of his mouth like a torrent.

He first offered a brief but eloquent commentary about how he and John had shared certain prophecies but took different stances on other issues. He then fell silent for a moment and the crowd wondered whether he had altogether lost his train of thought.

But then, Jesus pivoted toward Simon,
Who had already folded up his nets
    to place them to dry upon shore.
With an impassioned expression on his face
    and an urgent tone in his voice
        that alarmed the crowd,
Jesus commanded of Simon,
    "Fella, loosen your boat
and launch it
    into the deeper water."
He pointed to a place not far off,

where a rivulet

came into the sea

from a side canyon.

"NOW! Let down your nets

to prepare for a big catch."

As the crowd watched Jesus wildly gesture to Simon and his crew, they were tickled by the fact that a landlubber who had just walked in from the desert would be audacious enough to tell a seasoned angler where to fish, especially after all the boats on the water had come in empty that morning.

Simon, already so weary

that he could hardly

feel offended

or embarrassed, simply replied,

"We've been hard at it all night long,

and we didn't catch a damned thing.

But if you insist,

I'll go out with you

and I'll cast out the trammel nets

where you tell me to."

So they rowed the boat out to a place that Jesus had pointed out, a spot on the shoals along the northern shore of the lake, at the foot of a ridge above the waters.

It suddenly dawned on Simon that Jesus was leading them to a little-remembered but anciently esteemed fishing ground called Heptapegon, or "Seven Springs." It was a place the elderly fishermen who Simon apprenticed with in his youth had associated with the tilapia called *musht*. This prized fish gathered there only in the winter around the warmer, nutrient-rich freshwater upwelling not far offshore.

Because the musht could see the cords of trammel nets when they were used during the day, other forms of netting were typically preferred until dusk. In fact, trammels were best used on moonless nights.

Now, however, Jesus was beckoning them to use trammels after sunrise, which seemed nonsensical to the fishers, who had folded up their trammel nets as the first light appeared on the horizon. But for some reason, they obeyed the odd dry-lander, lowering the trammels and setting the bottle gourd floats to them in a way that would make them bob if the nets began to fill with fish.

When Simon tugged on a rope connected to the nets, it was heavier than he could pull up on his own. The net was filling up more rapidly than any they had tossed out in the water for months.

"Help me pull the nets up!"
    Simon called to Jesus
       and the rest of the men.
As the entire crew grabbed the ropes
    and the upper rungs
       of the nets,
They could see that the flaxen webbing
    had begun to tear
       under the weight
         of such an abundant catch.
Even though they were weary
       and had been up all night,

    They persisted in pulling in
     the entire webwork,
   now somewhat tattered,
     to place it back into the boat.
It nearly sank
    under the weight of the catch
with hundreds of two-pound tilapia
   writhing and leaping
     all around them.
The sight
   of the bounty
     humbled Simon.

A stranger who had come in from desert lands had forced Simon to remember this secret fishing ground, known as much for its fragility as for its hidden bounty. His father, Jonah, and old Zebedee had cautioned him to fish there only if their families were verging on starvation or bankruptcy. Now was one such time. Their livelihoods—if not their lives—had been saved by this mysterious visitor.

> Simon fell to his knees in front of Jesus,
>> amidst all the fish flopping
>>> and filling the hull of the boat,
>> he cried out to Jesus,
>>> weeping with joy and humility,
> "Have nothing to do with me, Master,
>> for I had somehow come to believe that God
> would never again bestow upon us
>> such abundance in this world."

Jesus quietly put one hand on Simon's shoulder and reached out his other to help lift the man up.

Simon, his brother (Andrew), and their hired crew brought Jesus and the heap of fish into shore, to the amazement of the crowd. Then James and John, the sons of Zebedee, brought their other boat over to meet them in order to transfer some fish out of the first, for it was stuffed up to its gunwales. They were stunned by the sheer bounty of fish, the likes of which would allow all the partners to pay off their debts and continue to fish together.

Simon looked at them, teary-eyed,
    unable to explain what had just happened.
Jesus simply said to them,
    "Do not be afraid.
It is clear that you know how to fish,
    but from now on,
        you will be fishers
            not of musht
                but of men."

Talking among themselves quietly, the anglers then announced to Jesus that they were ready to abandon everything else they had known in their lives to follow him, to listen and learn from this wayfaring stranger.

4

SUSTAINING ABUNDANCE

Whenever I hear this story about the abundant catch at Seven Springs, I am reminded of another story I heard from a friend who lived and worked in a traditional fishing community of Cree Indians on the James Bay in northern Canada. It was there that ecologist Fikret Berkes observed something that later startled scientists when he reported it. In an era when one fishery after another frittered away, the fish populations of James Bay were stable, for they were being stewarded differently—and more sustainably. In a later chapter, we'll see the Cree independently came upon some of the same fish conservation strategies Jesus espoused in one of his parables.

Cree fishing families had long been better at fishing sustainably than their nonindigenous neighbors. Those neighbors resolutely adhered to regulations mandated by Canadian law but still overharvested one fish resource after another. The Cree traditions kept the entire populations of two fish species from being overexploited, even though Cree practices ran counter to what professional fishery scientists were recommending to government regulators at the time.

The Cree removed the biggest specimens of cisco and other whitefish from the nets, kept them to eat, and celebrated them like a sacrament at their communal feasts. But they insisted on releasing most of the smaller-sized individuals caught of the same species so that they might continue to grow and ultimately breed, thereby replenishing the stock.

Rather than seeing their fishery depleted by high-grading off the largest fish, the Cree sustainably maintained the population by liberating most of the small fry rather than harvesting and eating as many as they could.

Such ethics of community self-regulation to avoid resource overexploitation were once widespread among subsistence fisherfolk all around the world. That's because these fishers lived and angled or netted in one place long enough to see the consequences of their own actions play out in the relative abundance of fish found from year to year. Instead of being held captive by greed—taking and selling every last fish that entered their nets and boats—they opted for self-restraint, a slow and steady way of fishing that sustained abundance.

We can only speculate that such conservation practices were once implemented among the fishers in the
Sea of Galilee. But we know that when the insatiable
markets of Rome increased outside demand for fish from
Galilee to be pickled, pasted, salted, or smoked for export,
something went awry.

Not many Biblical scholars pay attention to the
subsistence-oriented stewardship by the fellaheen and
the export-oriented extraction by foreign seafood brokers.
But issues like this were at the core of social conflicts when
Jesus sailed among the anglers of Galilee.

Look around you for a moment. Take a good whiff of the
fragrances in the air. Where do you perceive an abundance
of life-giving resources? Where do you sense that something essential is lacking, so much so that you can't see
how anyone could make a living off the meager resources
in full sight?

Have you had times in your life when you fully felt
what it is like to be poor? It's that fatalistic sense that your
pieces of the pie are getting smaller, so much so that you
cannot even imagine that the pie itself will ever get big-
ger. Whatever opportunities or resources those around

you harvest or capture seem to diminish whatever you can garner. In economic theory, this is called the zero-sum game. Someone garners all the winning cards so that you must fold your hand and toss in your last chips.

Well, Jesus never played the zero-sum game. He could always find abundance where others saw scarcity. He witnessed so many joys and laughed so hard among the poor that he could not for the life of him figure out why the elite thought they were impoverished!

K. C. Hanson was painfully aware of the disparities in access to resources during Jesus's time, but Hanson also saw the thread of hidden abundance weaving through the parables. Hanson is a scholar who has explored the ecological dimensions of the parables about fish. He thinks like a fisher as much or more than he thinks like a philologist, philosopher, or folklorist. For decades, he resided along the Pacific Seaboard, but earlier in his career, he spent eight years in Minnesota, "the land of 10,000 lakes."

Maybe this period of his life gave him a particular understanding of the burdens of freshwater fishing in the Sea of Galilee. In *The Galilean Fishing Economy and the Jesus Tradition*, Hanson lamented that scholars haven't taken the time to explore Galilean fishing apart from the basic observation that four of Jesus's disciples are identified as fishers. But fishing would have had to be a huge part of daily life that indirectly influenced spiritual practice, since it was such a vital part of the region's contributions to the Roman Empire. It gave fishermen their vocabulary for expressing their relationship with the world.

What kind of fish are we talking about? We've already met *Oreochromis aurea*, the tilapia we've been calling "musht," which Christians later nicknamed "Saint Peter's fish." The Sea of Galilee still harbors eighteen other native fish and at least nine introduced ones. Among the natives, eight species in addition to the musht belong in the cichlid family, including four other tilapia and two freshwater "sardines." All in all, dozens of tons of these sardine-like fish were annually harvested during the life and times of Jesus.

A single carp-like fish called a *blenny* is still sought after in this freshwater sea, for it is one that has long been popular on the Sabbath and at wedding feasts. The Sea of Galilee also has an Asian killy (or Persian pupfish) left, one air-breathing catfish (which is edible but not kosher), two loaches, and a few carpels and bleaks.

These fish were skillfully pursued by Galileans, providing a significant commercial fishery for them over the span of more than 2,500 years. The fish were caught with dragnets, seine nets, trammel nets, cast nets, wicker baskets, fish traps of woven branches and reeds, spears, arrows, pronged tridents, hooks, lines, and sinkers.

The catches brought in by families on the Galilean shores were often clandestinely sold, gifted, or bartered on beaches where a boat could moor and a quick exchange could occur. In that manner, the fishers could evade the tax collectors who wanted to take part of their catch. The empire mandated that taxable catches destined for export markets must be taken to a single market in a highly prosperous but corrupt village called *Magdala*. Yes, that's where Mary was from—a place famous for its inequities.

There, the fish were purchased wholesale, salted, and sun-dried, then made into a salty slurry or "pickled" so they could be transported long distances to other hubs of commerce in the Roman Empire. Some of the "pickled" fish and fermented fish slurry were further processed into the iconic fish paste called *garum*. Clay pots full of garum fish paste could be readily transported all the way to Rome, where it was in great demand.

That was the trouble. Ever since the sea itself had come under the control of the distant empire of the Romans, ordinary fishers as well as the fish stocks they depended on had become impoverished. Hanson says their boats were regulated by the state so that the elite could profit from their labor. Aristocratic families controlled the roads and bridges that fishers used to transport their catch to market, and the duty rates could be up to 5 percent. That was in addition to the bribes and tithes they had to pay to the Herodians who ran the harbors.

These external economic pressures likely forced the anglers to stay out longer in the water and to use more predatory or extractive methods of fishing that threatened the sustainability of the Galilee fisheries. What else could they do? Just to survive and support their families in Galilee, fishers there worked harder and longer merely to stay in the same place.

This is perhaps the greatest injustice that imperial powers can impose on fishers and farmers whose lives depend on the stewardship of resources: they force the fellaheen to squander the fisheries' resources they rely on for their own sustenance. Soon, the most prized fish are priced out of reach for eating by the very families that brought them out of the sea.

Their burden of debts, taxes, and tariffs pressed them to harvest more fish than what their own ancestors would have ever considered as "an allowable level of take."

The exploitation of fishers in Jesus's day is still true of the Galilee today. Fish stocks are also imperiled by the contamination of water: the dumping of manures, toxins used in tanning and from industrial processing, and human wastes from households. The uneducated poor are often indicted as the perpetrators of such pollution. But they typically cut corners and relax ethical standards when their backs are up against the wall and they have few economic options left.

Many native species have gone extinct in the Sea of Galilee since the time of Jesus. Why? Well, for starters, both the quantity and quality of its water have so dramatically declined over the centuries that many fish got left high and dry. That loss of habitat has eliminated the sacred *Tristramella* fish and a small *Daphnia* crustacean from the lake. They are gone, never to return.

Perhaps the saddest demise of a fish population in the Sea of Galilee has happened in recent years. A precipitous drop in numbers has imperiled the very tilapia linked by legend to Simon Peter, the same fish we saw in the story about Simon's call to be a disciple.

Around the turn of the new millennium, a fisher named Menachem Lev spotted one of these tilapia that was unusually thin. One of its eyes had popped out of its socket.

He soon noticed that other fish were blackened and covered with red spots. These fish had lost sight in both eyes and had difficulty avoiding any hazards they might have faced. Unable to forage because of their blindness, they began starving to death.

Lev began to sort the blinded fish out from still-healthy ones that he sold to restaurants around his home in Kibbutz Ein Gev. He took the sickened fish to Israeli laboratories for analysis, where pathologists determined that the blind "Saint Peter's fish" were fatally infected with a previously unknown nerve virus that caused necrosis of the flesh. By 2009, it was clear that this lethal nerve disease had infected at least a tenth of the Saint Peter's fish remaining in the sea. They were dying by the thousands.

In 2010, Israel's ministries had to ban all fishing in the Sea of Galilee, temporarily suspending a livelihood that had persisted for millennia. All commercial stocks had fallen to such dangerous lows that several species were being considered for listing as globally endangered.

Lev and the other commercial anglers of Galilee found themselves with no nightly catch to bring home, some two thousand years after Simon Peter and his crew had briefly experienced the same disappointment.

The aqueous world beneath their feet changed irrevocably. For Jews like Lev, as well as Christians and Muslims who pledge to care for all of creation, there is a responsibility to help not just the blind and the hungry within our own species but also the blind and the hungry within other species.

Sustainable abundance is possible in our waters. The fishing parables of Jesus did not explicitly anticipate such modern-day horrors as what has happened to Saint Peter's fish in our day, but his words clearly recognized that a sea was being depleted and its fishers impoverished. And yet Jesus did not sound at all like one of the doomsday zealots of his times. Most of his fishing stories were filled to the brim with hope, calling us to move forward with contemplative and compassionate actions that replenish our seas and restore our soulful communities.

5

CHANGING THE WAY WE FISH

Who among us was not enthralled as a child by landing a fish that ended up wriggling in our hands before it made its way back into the water or to the dinner table? It may have been the first time that we were confronted by the awesome paradox of food getting—that another life is taken to nourish our own. We must later learn to do so prayerfully, compassionately, and sustainably or we risk imperiling the very plant and animal populations that our lives—and the lives of future generations—depend on.

In the spirit of sustainability and compassion and with a view to the future, Jesus told a parable from the edge of a boat not long after he told his friends the story of the sower. As Matthew remembered it, Jesus was sitting beside the lake, maybe soaking his feet in the shallows to ease the soreness in them. Before he knew it, a large crowd had gathered around him, hinting that they wanted to hear another story, crowding in upon him so that everyone could hear. He crawled up into the nearest boat so he could look into the eyes of everyone who had gathered. He sat there for a while in silence, making eye contact with each and every person.

Suddenly, the words began to pour out of him like a gushing spring as he offered another parable to all the people who stood enraptured on the shore:

> "Again, I ask you, what would you say
>> our circle of kinship should be like?"

He let them think for a moment, then he began again:

> "If you ask me my vision for that heavenly kin-dom
>> it will be like a fishing net thrown into the water
>>> that miraculously gathers up
>>>> fish of every kind!

> But when a net is that full,
>> the fishers have to do some winnowing.
>>> So they drag the net
>>>> full of wriggling fish
>>>>> onto the shore, where they
>>>>>> sit down among the harvest

to begin to select the fish

     mature enough to keep

while tossing the immature ones

     back into the waters

to grow some more,

     for they are simply not ready.

That is the way it may be at the end of this era.

     The sacred messengers will come to gather

     all who are ready

while deciding which of the others

     unsure of themselves

     must be tossed back

     into the perilous waters.

My heart will go out to you when

     our harvest is sorted out,

for we may hear weeping

     and the gnashing of teeth,

for some may live on with us,

     while others may perish.

but do you understand why

     I wish to caution you?

We must prepare

     to take care of this bounty."

My own family's history has "schools of fish running through it" as a passion and pleasure, but we also witnessed firsthand the decline in fishing livelihoods in our community. While my grandfather "Papa" Ferhat Nabhan was not a fisherman himself, he settled among Swedish fishers when he moved his family from Lebanon to the Great Lakes a century ago. When he was a middle-aged man, he made his living as a produce peddler, but he regularly bartered fruit for fish with his tow-headed neighbors at harbors along the southern shores of Lake Michigan.

I've met elderly Swedes who claim that Papa kept their families from starving during the Great Depression. He gifted them all the fruits they could eat while paying them well for any yellow perch they had caught so that he could resell them to his more wealthy and elite customers.

But sometime in the 1930s, a predatory fish named the sea lamprey was accidentally introduced into the Great Lakes, as was a small oily fish named the alewife. Their populations grew exponentially, especially those of the lamprey, which ate forty to fifty pounds of native fish over its lifetime. Soon, the native stocks of perch, sturgeon, and cisco were so depleted along the shores of the Indiana Dunes that all but one Swedish fishing family abandoned their shacks and boats by the time I was born there.

After the sea lamprey came to dominate the lake's food chain, the alewife population peaked and then collapsed—with millions of stinking carcasses washing up on the lakeshore. No Swedish American family of commercial fishers could make a living on the southern shores of Lake Michigan anymore.

What happened in my family's lakeside village is just one example of a crisis emerging along many shores. Today, we are witnessing a global crisis in our fisheries, one affecting artisanal fishers living along the shores of nearly every sizeable freshwater lake in the world. For many, this crisis still remains hidden, but it's real. The "empire" we live within is no less extractive and destructive of creatures like fish than the Roman Empire that ruled over Galilee.

Today, four out of every ten species of freshwater fish left swimming in the lakes and rivers of the North American continent are facing some level of endangerment. If Simon Peter and his crew were alive today and trying to fish in the freshwater lakes of the US, would they be able to make a living?

Some, like the Atlantic sturgeon found in Eastern rivers, once reached sizes of eight hundred pounds. Their swimming power had the sheer force sufficient to tear through fishing nets as if they were tissue paper.

As early as the 1920s, the populations of sturgeon in one river after another blinked out, diminished by overfishing. I have only held two fingerling sturgeons in my hands over decades of visiting Eastern rivers and every Great Lake. As I tossed these survivors back into the waters, I wept and prayed that they might find safe passage away from nets and hooks.

But that is not the half of it.

Most of the once-populous fish in North America have been imperiled by many factors other than the actions of fishers themselves. We have witnessed whitefish and yellow perch precipitously decline by the damming of rivers and draining of wetlands; by the warming and drying weather that depletes water supplies; by the draining of toxins and endocrine-disrupting chemicals from our sewers and crop-lands into our waterways; and by the spreading of aquatic weeds, diseases, competitors, predators, and pests that were previously unknown.

Government regulators claim that they have made advances in fishery management and conservation over the past hundred years. That is indeed true to some level. Most of these regulators are well trained and care deeply about the resources they seek to manage.

Nevertheless, both fish and their fishers are now facing a crisis at least as severe as the one faced by sardines, tilapias, pupfish, bleaks, barbels, and Galileans during the brief fishing careers of Jesus's disciples.

However many fish stocks we have safeguarded through the more scientific management of fisheries, we have lost far more through wastefulness, greed, and disregard for the creatures and characters whose lives depend on clean waters. As Jesus's parable warned, the empire of extraction is swallowing up everything that breathes.

Many Christians today don't seem too concerned—or are barely aware—that many fish stocks today can no longer make their annual "runs" or migrations that have gone on for millennia. They might ask, What does that have to do with our own spiritual pilgrimages? As if we are the only species that makes sacred journeys in this world!

As theologian Diane Butler Bass reminds us, most ministers routinely use the image of sacred waters in their preaching. But they make barely a single passing note of the global fishing crisis occurring in our rivers, inland lakes, and seas. She says we can do better because "the world's waterways call us to practice social justice—to restore them, to make sure rich and poor alike have access, and to manage water in drought-stricken lands with creativity and foresight."

We have an opportunity to make a difference and change how we regard Brother Tilapia and Sister Salmon. And there's no time to lose. According to British fisheries journalist Charles Clover—whom I met on a panel at the international Slow Fish Expo—"we're at the end of the line." If we don't stop squandering fish stocks, millions of people could starve. Pavan Sukhdev, of the UN initiative for greener economies, has warned that if there is not a fundamental restructuring of the global fish industry, we may encounter fish-free oceans and lakes by 2050.

What is a world with such a dearth of fish that children will grow up without ever having felt one wriggling in their hands? One where fish are *absent* from our lives?

The health of at least a billion people, mostly from poorer countries, who nutritionally depend on fish and shellfish as their main animal protein sources would be at risk. And that's why efforts to restore their coastal, estuarine, and riverine nursery grounds give me so much hope. Where my wife and I spend time on the Gulf of California coast of Mexico, we are assisting the Comcáac Indian fishing communities in restoring mangrove lagoons and seagrass beds. That's where dozens of finfish and shellfish mate, have babies, and feed. A single hectare of restored healthy mangrove habitat—hardly two and a half acres—brings $30,000 of edible fish and shellfish back to Gulf waters, in addition to hundreds of other nonedible but ecologically important species. The replanting of mangroves and eelgrass beds also creates jobs for youth. It is one of many means to offer long-term care for the bounty of fish.

Just as Jesus found fish in a place where his Galilean friends had given up thinking they could ever bring in an ample catch again, all of us can pray and work toward restoring abundant harvests wherever we live.

6

HIDDEN TREASURES

The collapse of fisheries has triggered spiritual as well as economic consequences for fishers. To reflect upon this dilemma more deeply, let's take a closer look at the Saint Peter's fish we've already talked about. It has some unique behaviors that Jesus must have recognized.

Saint Peter's fish is one of several kinds of tilapia that functions as a "mouth-brooder," a behavior that ancient Galilean fishers knew of well before scientists described it. We can tell that this was common knowledge among fisherfolk in Galilee through the details surrounding the "coin in the fish" miracle attributed to Jesus in Matthew 17:27. As fishers and their families cleaned and gutted tilapia on the shores of the Sea of Galilee, they were especially curious about the stones, coins, and other small objects they found inside their catch. Jesus himself must have known at least this much in order to spin out one of my favorite parables.

During their reproductive period in the springtime, females of this kind of tilapia may spew out their masses of eggs in a pit excavated by the males on the lake bottom, where the males then fertilize them. Each mother then gathers up her eggs by nudging them into her mouth and passing them back into her buccal cavity, where they incubate for one to two weeks.

The female will not ingest any food while she has the new hatchlings still in her mouth. Once they are big enough to release, she regurgitates them out into the water.

The newborns remain so excitable and fearful that for several more days, they will try to swim back through the mother's lips. Fishers claim that to prevent the small fry's reentry, the mother fish will pick up pebbles, coins, or other small items while foraging on the lake bottom to place in their mouths and buccal cavities as obstructions. For centuries, fishers who have caught mouth-brooder tilapia in the Sea of Galilee have encountered such items in the fishes' cavities.

This is not rarified, esoteric knowledge restricted to only a few specialists. It fascinated many who have lived on the shores of Galilee over the centuries. As they brought in fish, they would check their mouths and cavities for "hidden treasures."

If they discovered an object that a tilapia had gathered and sequestered, they took it back into the village and presented it at the nightly social gathering called *haflat samar*. A haflat samar was an event where oral histories were told to preserve the traditional ecological and social knowledge of the community. Here retired fishermen could recall and recite all the objects redeemed from Saint Peter's fish over their lifetimes.

Think of all these delicious bits of social and natural history as you see Jesus, Simon Peter, and the other disciples come into Capernaum, the most prosperous fish market in Galilee. The Bible says that when they arrived in Capernaum, Simon was confronted by bureaucrats who made their living by collecting a two-drachma temple tax. They knew that he had not been coming to the temple to hear the rabbis there but had been listening to Jesus on the streets instead. They slyly challenged him by asking,

> Your teacher must certainly be willing
> to pay us his tax,
> won't he?

The disciples looked at one another, incredulous. It seemed ridiculous that a street preacher from afar had to pay a tax simply to pray or preach outside the synagogue. But Simon wished to avoid any conflict, so he tried to placate them, replying, "Of course he will."

But when he arrived back at his home where Jesus was resting, Jesus sensed the anxiety in Simon's heart, asking what he was concerned about. When Simon told him of the confrontation, Jesus presented him with an odd request:

> Go down to the sea, cast out your line or your net,
> and wait until the fish rise,
> then haul them in.
> Take the very first fish you catch,
> open its mouth, and inside its cavity,
> you will find a four-drachma coin.
> Take it to the tax collectors,
> and please use it to pay for both of us.

This request from his teacher made no sense to Simon. Jesus was clearly not opposed to paying the tax collector, but why would he want to get the money to do so from a fish?

Jesus used such a nonconfrontational strategy with power-mongers over and over again. He aimed to deflect the force of oppressive acts perpetrated by secular authorities against farmers, fishers, and wayfaring strangers.

Instead of willfully arguing with the tax collector and protesting the political and economic powers that the tax represented, Jesus responded from a higher level of consciousness. In this case, he proposed using "a gift of nature" to satisfy the petty demands that the temple tax collectors aimed at him and Simon.

But the request to placate the tax collectors by obtaining a four-drachma coin from a fish must have sounded preposterous to all who were present. It also seemed demeaning. Already humiliated as poor peasants without any money to spare, were they now to go begging for coins from a fish?

But then, perhaps something rose up in Simon's memory that he had not considered when Jesus first made his request. We can picture Simon remembering how his father and his friend, old Zebedee, had once told him about finding a precious coin in the cavity of a freshly caught musht. Once again, how did this landlubber teacher know more about the hidden gifts of the sea than most fishers around him did? Even as a man descended from a long line of traditional fishers, Simon had to be nudged by Jesus to see a gift from the sea that was already within his reach!

Jesus had the astonishing capacity to remind all of us of the presence of precious but hidden treasures in the natural world. These gifts are already in our midst but are often ignored or dismissed by us.

The natural and so-called supernatural are not separate layers in some parfait glass—they are inseparable.

By simply challenging our perceptions, as Jesus and other great teachers have urged us to do, they have opened us to the grace that is already all around and inside of all of us. Imagine that: a coin inside a fish can be miraculously summoned to solve a seemingly insurmountable problem of paying off a debt when you are dirt broke!

Jesus and Simon could cleverly meet the bureaucrats' preposterous demand with an altogether improbable—if not wondrous—solution.

In doing so, they put their faith in the abundance found in creation itself, the very expression of their Creator's bounty that was everywhere around them. It transcended the man-made wealth of the Roman economy. When we least expect it, when the odds are statistically against it appearing, it's there when we most need it: the hidden treasure of the bounty of creation.

The miracle was not simply about finding a tilapia with a much-needed coin in its mouth. The miracle is about breaking out of our blinders and seeing God's grace wherever we go, regardless of the challenges placed before us.

FARM STORIES

Jesus's respect, affection, and compassion for the sad flock of imperiled farmers and fishers of his time meant that he did not sugarcoat their prospects for a dignified future, as if he himself could assure success for them.

And it will do us no good to sugarcoat the challenges that agrarians are facing today, including rural poverty, rising suicides, and a poisoned earth.

Yet a bountiful harvest is still achievable by anyone who tries to make the earth good by their work. You yourself may already be playing a role in such blessed work. That is why we need to keep your spirits up, just as Jesus tried to do with the farmers of his era.

7

IT IS HIGH NOON
IN THE DESERT

Have you ever been in the desert at high noon when the unrelenting heat of summer shimmers in waves above the ground's surface? Have you ever tried working under such arid conditions?

Did the dryness parch your throat and your lips? Did the heat and glare of the sun make your head ache or your skin feel as though it were burning?

Imagine how such heat stress forced farmworkers to flee from their fields. In the old days in Galilee, they might take a break from their labors to head over to a grove of date palms. On the grove's edge, they could sit in the shade of a few feral fig trees.

Those trees grew from seeds carried, defecated, and planted by Egyptian fruit bats and birds up on the rocky ledges of the steep cliff face. Climbing beneath them were native grapevines and bramble bushes sprawling along a small rivulet trickling into the lake.

Along the shores of Galilee, such a grove was among the few places that had adequate shade in which to rest. What made it even better was the access to delicious fruits and wild berries that belonged to no one and by the fact alone were available to anyone.

That is the setting for the story Jesus is about to tell.

The men are busy slaking their thirst from jugs of water and nibbling on crumbs of day-old bread, smoked fish, and figs. They do not initially notice that a stranger has appeared in their midst, accompanied by a few of their compatriots, ones with whom they fished or did farm work in their youth.

At first glance, they pay the stranger no mind, for they simply banter with their old acquaintances.

About that time, a young boy—the son of a fisher—walks by him with a basket full of dried figs on his head. These are wild figs that he and his grandfather have picked from the shorter trees that grow in a dense thicket beneath the palms. The grandfather has grown too old, scarred, and crippled to be of much help to his sons out at sea.

The boy hopes that someday he can join his father and uncles in their work. That is, if they can ever pay off the debts they have accrued to hold on to their boat. When his father returns, his nets are nearly empty except for a few small fish. But when he sees his son with the basket of freshly picked fruit, he beckons him to share the harvest with the crowd resting in the shade.

Before the first hungry person can be fed, Jesus grabs the basket full of figs off the head of the child, who looks up, stunned with bewilderment.

Jesus swings around with the basket of figs, his arms stretched out before him to offer a fig to each of the tired, hungry workers. The crowd laughs at his theatrics. He has caught their attention, and now he sets the stage for a story, continuing to circulate among them, giving each of them a fig.

He disappears into the thicket below the date palms. A few moments later, he emerges and returns to the crowd with the sinuous leaves of a wild fig tree, wild grape leaves, the hooked thorns of a holy bramble—all in one hand—and a perfectly ripe purple fig held up high in the other.

"You will know *each tree by its fruit*," he begins slowly. Of course, they all already know this, but they play along as he walks around with his botanical specimens.

Are such figs gathered from a mess of thorns?
Can grapes be picked from such a bramble bush?

They guffaw, since the answer is obvious to anyone except a city slicker. So Jesus points up to the fig tree towering above them, with its deep-green, glossy leaves that manifest a sign of wild health; its roots reach down into the water of some hidden spring.

Then he points to the dry streambed beyond, where the insect-infested leaves of the native grapevines and bramble bushes growing in the dry gravel look sickly by comparison.

No fig tree as beautiful and vigorous
    as this produces sickly fruit.
Nor do sickly vines or bushes produce
    fruits as luscious as these figs.
If he is fortunate enough to realize the abundance
    that is right before him,
A good man will produce more good in the world.
But if a discontented man acts
    in a sick and selfish manner,
        only bad will befall him.

He stops talking for a moment, and the men do not stir, each silently asking himself what kind of tree he might be.

Jesus pops one of the ripest figs into his own mouth and slowly savors it. He thanks the boy and returns the basket, which still has plenty of fruit within it. He then whispers so softly that not everyone present can hear him:

You know, your personal fate is not already set:
you can choose which kind of plant you wish to be.
But it is only out of a generosity of heart
that an abundance of flowers and fruit can flourish!

That query from Jesus has echoed across the ages: Just what kind of person—farmer or fisher, father or mother, leader or follower—does each of us wish to be? How different is that person from the persona or masks others wish us to wear? We must "decolonize" ourselves, just as Jesus did when leaving his family's preferred work as a *tektōn* to move among fishers and farmers as a street preacher of immense gifts. Whether we grow a tiny garden or a thousand acres of corn is not the issue. It is whether we do it with the passion and compassion of a true calling. As poet Mary Oliver echoed Jesus's question two millennia later in her little gem "The Summer Day," "Tell me, what is it you plan to do with your one wild and precious life?"

8

FARMSTEADS, HOUSEHOLDS, AND COLLABORATION

Jesus placed a fresh lens before the eyes of the down-trodden. To be sure, he was clearly concerned with the social and political injustices being suffered by the fellaheen. He could see them suffering dire consequences with nearly every step he took along the dusty roads and donkey trails of Galilee. But if he wished to change the dreadful power dynamics that were afflicting many of his farming friends living in that landscape, he knew it would not be accomplished merely by raging against the powerful. The poor needed hope to crawl out of their economic ravine. Many of them were sharecroppers, just like many Black families in the South during the Jim Crow era.

If he simply organized farmworkers for protests, boy-cotts, and acts of civil disobedience, they would have been crushed like so many grapes. Indeed, many of the zealots of his era were captured, mocked, jailed, beheaded, or crucified.

Some of the surviving but still wounded warriors went underground, but they often became listless, fearful, or prone to fight among themselves. Self-doubt and paranoia set in.

They not only felt terrified of their oppressors; they also felt terrible about themselves. Theologian Harvey Cox has aptly articulated how people feel and behave when they sense that the rest of their lives will be out of their own control as they struggle and compete for increasingly scarce resources: "We don't just live in the Empire. The Empire lives in us."

For Jesus to change the relationship between the peas-antry and the aristocracy, he had to give back to farmers their dignity and capacity to anticipate better days. He had to move them out of the prisoner's mentality of feeling that they had no viable options for improving their lot or their families' well-being.

Jesus had to convince them that they lived in a place where they still had access to the vitality of creation and the grace of the Creator. Instead of trying to compete with others to extract their "slice" of the ever-shrinking pie, thereby diminishing it further, they had to be convinced that respectful collaboration could help them regenerate needed nourishment.

Jesus had to persuade the downtrodden that there was an equitable and just "kin-dom" that lived within them, one more powerful than the Empire. In Greek, such a kin-dom was likened to a beautifully functioning and welcoming "household" or "farmstead," called an *oikos*.

That's the root concept embedded in our term *ecumenical*, "all being members of the same household." We also see it in *ecology*, which is "the logic or natural order of the earth as a multi-species household," and in *economy*, "the natural laws, ethics or conventions for managing a household."

As we'll hear in the following parable about some laborers in a vineyard, a farmer or landowner was called an *oikodespotes*, "the despot or steward of the household." Let's see how Jesus framed the ethical relationship between farmworkers and landowners.

The story begins as Jesus arrives one hot summer day on the edge of a farming village. There, he encounters some unemployed farmworkers resting in the shade. By the way they talk with one another, he senses that they are envious of the other laborers who have already been chosen for work that day. They felt overlooked and feared having to go back to their families empty-handed.

Jesus has a sack of freshly picked clusters of grapes slung across his shoulder. He opens it up and begins passing the grapes around for the workers to eat, for some are both hungry and thirsty.

Several of the men are content with this unexpected gift, slowly savoring each sweet grape. But one man spits them out, complaining,

> All the grapes you gave me are sour!
>> Give me some sweet ones
> like the ones others here with me were given!

That's when Jesus gives the rest of the sack to the whiner as he winks at the other men. Facing them all, he starts to tell them a story.

> The kin-dom of the Creator
>> works like this:
> Just after dawn on a hot summer morning,
> the master of a household went out into the street
>> To search for farmworkers

to bring in all the grapes from his vineyard.
After finding laborers for hire,
    at their usual gathering place
        on the edge of town,
    he promised this crew
        one denarius each
        for a day's work—
enough to feed a family for several days—
if they hastened to come with him and
    to get right on the work at hand.

The workers who are listening to Jesus see nothing out of the ordinary with this story; its occurrences are so familiar to them that some begin to daydream.

But the work of bringing in the vintage was overwhelming,
    so much so that within three hours,
he returned to town to hire more men,
    also promising all of them just wages
    if they would only promise to mobilize quickly.
    The sun beat down on the vines
and the grapes began to pop with the heat,
    as bees and other insects swarmed in on them.
By noon, the landowner had to return
    to seek out others for the hiring.

Finding a few more men clustered in the shade,
  he hired them right then and there.

His audience begins to perk up here, a bit irritated by the actions of this landowner who seems so clueless about what size of workforce he needs to bring the vintage in.

He did so again at the ninth,
  then one last time at the eleventh hour,
    which is when he asked the hangers-on,
      "Why have you held out hope to the very end
      rather than going home?"
They replied, "No one else has come to hire us."
  "Never you mind," he responded, "Just come with me."

This boss is so erratic, who would want to work for him? the day laborers in the audience ask themselves.

But the additional workers came in the nick of time,
helping to assure that the entire vintage was put up
  well before the sun went down.
Not long after evening came,
  the one who lorded over the vineyard reappeared,
    beckoning the crew boss to him.
He whispered as quietly as he could,

"Call in the workers
  then pay each set of farmworkers
      the very same amount,
  from the last ones hired to the first ones."

This gets their attention. What the hell do this master and crew boss think they are doing? The listeners are as stunned as the workers in the story, who each get a full day's wages, starting with those who only came in at the eleventh hour.

When those who selected to work at dawn
      saw what they had garnered—
      no more or less than the others—
  they pushed aside the crew boss,
      storming over to the landowner.
  They had much to grumble about,
      for they weren't given a larger share at all,
      gaining only what the others gained,
      even though the latecomers
      had barely worked a solid hour,
  while they themselves had slaved
      all day in the scorching heat.
  The landowner listened and then he replied
      to the one who whined the loudest,
          "Did I do you any wrong?

Did we not reach agreement
  at the start of the morning's labor?
Did I not say what wage would be fair,
  should you work for me the entire day?
Take what rightfully belongs to you, then go take a rest,
    for those who arrived to spell you late in the day
    can be offered as much as the first or best.
Am I not allowed to celebrate this bountiful harvest
  by offering its gift to whomever I choose?"

The audience, hearing this, doesn't make a sound. Some are humiliated by their own greediness and their lack of empathy for those who had waited silently without much hope. Although they did not know they would be hired, they then rallied to help the weariest in their midst.

They hear Jesus tell them to put aside their jealousy, to join in the generosity:

For this will not be the only time
  that the last will be put first
    and the first put last.

With disorienting twists and turns, this parable forces both landowners and farmworkers to consider what justice looks like for all players.

The generosity of the *oikodespotes*—the householder or vineyard owner—in justly rewarding all who helped him bring in the harvest. He challenges and may inspires all landowners to do the same for their workers. Amy-Jill Levine suggests that this parable is not exclusively about "good news to the poor," for it also addresses the "responsibility of the rich." It isn't simply about giving a token handout to the poor. It is about hiring and valuing people to do work in ways that can help them regain or sustain their dignity.

But this story also challenges the workers themselves— unionized or not—to get beyond petty jealousies. Collaboration is the key to rewarding employment and wages sufficient to ensure that everyone's families are fed. The workers and the owners need each other for the entire oikos, or farmstead, to function well.

This is what Jesus affirmed when he introduced the parable by saying that "*the kin-dom of the Creator should work like this!*"

9

FARMWORKERS

worked for a while as a day laborer in strawberry fields and apple orchards between a bout in a sewage treatment plant and another on a railroad crew of "gandy dancers." The older men who mentored and humored me were Cubans, Mexican and Greek immigrants, Puerto Ricans, and African Americans from the Mississippi Delta. Of course, some of them would never leave such "menial" jobs of grueling labor as I was later able to do.

Years later in central Italy, I briefly joined in the *vendemmia* by picking grapes destined for thousand-year-old-stone fermenting vats below a hilltop village. Our hands and arms were sticky with sweat and grape juice by the end of a seven-hour day.

I have sown wheat behind draft horses pulling a moldboard plow, and at the season's end, cut the same crop by hand with sickles in the Tohono O'odham Indian village of Big Fields, Arizona. The elders who taught me to work with draft animals were the last of their generation in any direction for fifty miles to master the art of farming with draft animals.

While attempting to accomplish such fieldwork, I have been drenched by humidity, swarmed by insects, and felt the itchiness of dust and chaff on sunburned skin. I have suffered from brain and spinal injuries after being thrown from mules and horses. Like many rural dwellers, I have been gifted with scars and broken bones that have healed only years after the fieldwork was done.

But I have been fortunate enough to be able to step out of most kinds of sobering, muscle-straining chores as I aged past sixty-five; there are millions who cannot.

Consider those who harvest the grains for our daily bread and the fruit for our wine.

We once called them field hands, roustabouts, cowpokes, or casual workers, as if they are unskilled and inattentive, as if lacking in aspirations, brains, or proper names. Only belatedly—as a global pandemic threw our food supply into chaos—have we belatedly honored their problem-solving skills and diligence by deeming them "essential workers." As you read this from your easy chair or your tractor seat, they are cutting, reaping, binding, and threshing wheat with sickles, scythes, and winnowing baskets or with stationary threshers and high-tech mobile combines.

They're out there harvesting grapes by hand, pruning their vines with knives, clippers, and shears. They're irrigating field crops with shovels and hoses to guide rivulets of water into furrows, with drip emitters fed by PVC pipe, or with center-pivot sprinklers guided by computers and satellite imagery.

Climbing up into the canopies of apple trees on teetering ladders. Rolling out sheets of spun-bonded landscape fabrics to control weed growth between rows of tomatoes.

Herding cows with an all-terrain vehicle. Spreading manure with pitchforks and hoes; or applying anhydrous ammonia through mechanized irrigation systems. Slaughtering chickens with a prayer and a knife or using stun guns and gas chambers pumped full of mixtures of argon, nitrogen, or other inert anoxic vapors.

There are all manner of farmworkers who must go on, day after day, season after season, until they get too old or too hurt to continue: African American sweet potato diggers, Puerto Rican tomato pickers, Mexican asparagus cutters, Somali chicken factory hands, Hmong and Lao mushroom harvesters.

Some of them are out there in the North American landscape twelve hours a day, whether we notice them or not. Most urban Americans never catch sight of them, since farmworkers now compose only one-third of a percent of the entire US population.

Most Americans don't see the women and girls working in the rows, side by side with the men. One of those girls was Norma Flores Lopez, who began to work in apple orchards at age eight.

Norma worked four more years before she was able to collect her own paycheck at age twelve. Miraculously, she later went to college and became the director of the Children in the Fields Campaign of the Association of Farm Worker Opportunity Programs.

Despite her advocacy work and that of others, farmworkers are typically paid at the "piece rate," with wages based not on how many hours are worked, but on how many buckets or bags they pick of whatever crop they harvest. They may work in the field right through their pregnancies or other life transitions, simply to keep food on the table.

Today, at least seven out of every ten farmworkers laboring in the US are foreign-born. Many of them came to North America as refugees from warfare, political unrest, economic crises, or climate disasters that were disrupting their livelihoods in distant homelands. Where I live on the US/Mexico border, as many as ten thousand speakers of Spanish or indigenous languages come into Arizona each day of the harvest season, doing work in lettuce and onion fields that few American citizens wish to do week after week, year after year.

Once on American soil, they often take on the most hazardous tasks in the country. The death rate among farmworkers while on the job is roughly five times that of the average for all American occupations. Their vulnerability during the COVID-19 pandemic was off the charts.

Of the many ways that Matthew 25:40 has been translated into English, I prefer one that adheres closely to the original syntax in Aramaic:

> Amen, I say to you, as much as you have done
> to one of the least of my little brothers,
> you have done that to me.

Who of us regularly looks out for the "least" among us, those unseen workers whose lives support ours? One who took on the task was César Estrada Chávez. In over forty-two years of farmwork and activism in the borderlands, Chávez viscerally came to understand both what our society had done and what it had not done for the lives of field hands, day laborers, and their families.

Chávez began his activism as a civil rights organizer with the Community Service Organization in 1952; within six years, he had become its national director. Four years later, he and the equally charismatic Dolores Huerta founded the National Farm Workers Association, later called the United Farm Workers.

After Chávez's death in 1993, Huerta continued to raise the issue of farmworkers' plight for two more decades, sharing the perspective both of them gained about the place of farmworkers in American society. They are, she said, responsible for every stage of cultivating and harvesting our food—the greatest abundance of food society h-- -- But here's the tragic irony that she forced m. to face: "The food that overflows our market s our tables is harvested by men, women, and often cannot satisfy their own hunger."

Farmworkers are among the least recc paid, and least cared for contributors to A1 security and American culture. Black or brc still put down.

"The least" is not hyperbole. National Farm Worker Ministry (NFWM) has confirmed that agricultural workers such as field hands, fruit pickers, and herders have the lowest annual family incomes of any wage-earning professions in the US.

Many farmworkers prefer to be invisible to the government and even to community nonprofits offering them various kinds of health and hunger assistance—they are part of another vulnerable population, the "undocumented," who are subject to deportation. Many farmworkers have difficulty accessing health-care services or become confused when trying to navigate the programs for which they are eligible. According to the NFWM, this confusion leads to farmworkers "waiting until their health problems are unbearable" before seeing a doctor.

Some of those health problems are quite serious, resulting from excessive exposure to sun, heat, poor sanitary conditions, and pesticides. In a given year, as many as three hundred thousand farmworkers may become sickened by pesticides while in fields. Because farmworkers have high levels of exposure to pesticides both in the fields and in their field-side barracks, they have an unusually high vulnerability to Hodgkin's lymphoma, Alzheimer's, and persistent skin rashes.

It is not surprising that young men (four out of five farmworkers) and young women (one out of five) don't last long in this profession, given the physical dangers, the high death rates from accidents, the heat stress and toxins, the cumulative wear and tear on the body, as well as the hunger and poverty associated with rural jobs out of doors.

The Beatitudes offered as part of Jesus's Sermon on the Mount remind us that those who labor in the fields should continue to be of special concern to everyone, not just a few "farmworker activists" like Chávez and Huerta. Biblical scholar Ellen Davis notes that the word used for "poor" in this part of the Gospel of Luke is a special one. It's not just about lacking money; it is also about not having a voice or a place at the table. As such, Luke's version of the Beatitudes reads as if it could not be any better fitted to a segment of our society than it is to undocumented immigrant farmworkers:

> Blessed be all you poor ones:
>> for yours is the kin-dom of God.
> Blessed are you that hunger now:
>> for you shall be nourished.
> Blessed are you who weep now:
>> for you shall have a chance to laugh.
> Blessed are you, even when men shall hate you,
>> and try to separate you
>>> from their company,
> and to reproach you
>> and cast out your name as evil,
> as if all this is done for the sake of the Son of Man.

Among the ranks of farmworkers and day laborers are those migrants who move from place to place with the seasons, following the harvest. Jesus had something in common with migrant workers. His parents apparently took him with them on their flight to Egypt, where they may have toiled in fields along the Nile. He once likened himself to desert foxes who had no dens and to the birds of the air who had no nests. As an itinerant rabbi, he stayed for a night or two in the homes of his youngest disciples' parents or on a mat on the floor of the older fisherfolk he knew. Jesus knew what it meant to have "no home in this world anymore," as Dust Bowl songster Woody Guthrie once wailed.

There are many people among us who move from job to job, place to place, trying to patch a transient life together. They possess no property; they cannot be sure from one moment to the next that they will find some safe and dry sanctuary in which to lay their heads.

Since 1996, the number of American children living as vagrants in such extremely impoverished conditions has doubled. Most of their families have no safety nets. Barely a quarter of these poverty-stricken families received welfare assistance by 2013, down from two-thirds of the families in need two decades ago.

Seasonal migrants stay in a sequence of camps and cabins as they follow the annual cycle of harvests from region to region. They are caught on the treadmill of the food supply chain. Their incomes are often insufficient and too erratic to garner yearly leases with prospective landlords.

There are somewhere between three to five million American farmworkers whose relationships with other family members are disrupted by their roaming to follow the harvest.

They are "our" berry pickers, sheep shearers, vine grafters, bulb transplanters, cullers, potato diggers, gleaners, and mushroom hunters. They are our asparagus shoot snippers, fruit tree pruners, oyster spat seeders, testicle cutters, branders, vaccination injectors, ear taggers, hide tanners, offal collectors, crayfish catchers, and clam diggers.

Alternatively, we are their consumers. They do the "dirty work" for us: the labor of getting fresh, clean, processed food onto our plates.

You may recognize them by their work clothes, not by their individual faces, voices, or names. They wear gum boots, wet suits, chaps, bandanas, wide-brimmed hats, heavy gloves, dust masks, blood-stained aprons, twice-patched overalls, headphones, earplugs, nose plugs, and goggles.

Some young Latina women may labor all day long in layers of cloth, mosquito netting, and gear that conceal their countenances. I've seen busloads of Latinas heading out to handpick table grapes in fields where the air temperatures exceed 110 degrees and the ground temperatures reach even higher. Such temperatures can be life-threatening to those unable to escape: some experience shortness of breath, headaches, visual disorientation or hallucinations, accelerated heart rates, and muscle wasting. As a survivor of many heat waves in desert agricultural valleys, I am lucky. I have only been taken to the emergency room for heat exhaustion twice in my life. Tragically, such temperatures will soon be the norm in eleven western states as climate change makes working under such conditions unbearable. Workers will have to move on to greener pastures or perish.

In most states, more than half of the farmworkers and food processing workers are already migrants by one definition or another.

In the borderland county where I live, estimates of undocumented immigrants from Mexico run as high as eight out of ten of the Spanish speakers residing in our area. They conceal themselves in shabby motel rooms, dumpy double-wide trailers, and old barracks built for cowboys half a century ago. Such crowded quarters are like Petri dishes for the spread of illnesses like COVID-19 or influenza, as the 2020 virus hotspot near Chávez's birthplace of Yuma, Arizona, painfully illustrated.

They work a few months each year by sorting the glut of fruit coming across the border in the winter months.

They eke out a living the rest of the year by mending fences, stuffing and steaming tamales, mowing lawns, or hacking out firebreaks.

They seldom have the means to stay nourished very long. They survive only because an informal "hidden" economy still exists along the border that allows one family to take care of someone's livestock in exchange for a few months of shelter, or allows another to prune their neighbor's fruit trees in exchange for a few loaves of bread and a big bag of beans.

If we look at the Beatitudes again, it seems that Jesus was explicitly blessing and acknowledging the suffering of such migrants and castaways. What would his blessings sound like if he gave his sermon on the edge of our agricultural fields today?

> *Blessed are those who hunger and thirst*
> > *especially at the margins of the greatest surplus*
> > > *of food in the history of the world.*

> *Blessed are those who flip burgers in fast food drive-ins,*
> > *who fill burritos in taco trucks,*
> > > *or who go from house to house*
> > > > *with coolers of homemade goods*
> > > > > *hidden in their trunks.*

Blessed are those who patch together
their shelters from palettes and tarps.

Blessed are those who glean their meals
from the produce dumped on the edge of fields.

Blessed are those who are famished and fatigued
when nightfall comes.

Blessed are those who pray for Jesus to arrive
with fresh loaves and fishes.

Blessed are those who live with the fear that
Immigration and Naturalization Service will show up
with handcuffs and paddy wagons to deport them.

To bless them in such a manner, Jesus would have been
(and remains) with them. He still calls us to accompany
those who are the blessed as "the salt of earth."

INTO THE WILD

Have you ever noticed that what seems "hidden" before your eyes may say more about you than some elusive feature of the object itself?

The stories of Jesus may baffle us today with their paradoxical or sometimes arcane allusions. But while those parables may seem cryptic to us, they were probably clear to his contemporaries who did not have to work as hard to decipher them. That's because many of us have lost the traditional ecological knowledge of the wildness in the natural world, knowledge once commonplace in every farming and fishing community.

Jesus's basket of stories was filled with a practical understanding of rural life. This knowledge was critical in helping communities avert or survive crop infestations and fish die-offs, drought and flood, pestilence and plague, windstorms and heat waves. The parables that came out of his rural context are just as wise and timely for us today once we dig a little deeper into their meanings. Potential gifts emerge from the stories offered by the enigmatic but wise storyteller named Jesus of Nazareth.

10

THE WONDERS OF WEEDS

Maybe you remember this trope from a gardening talk you heard long ago: a weed is defined as "a plant out of place." Or maybe you've learned one pat answer in your experiences as a gardener while trying to grow your favorite vegetables over the years: weeds are bad.

Actually, a weed is defined not by an inherent hardwired proclivity for being bad or good, but by the context in which it grows. For example, darnel ryegrass plants found in fields are often weeded out because farmers see them contaminating, competing, or interfering with crop production. But the plant and its fungus can offer us both tangible benefits and intangible gifts if we understand them in a different light. To my surprise and delight, there are other contexts where their presence might clearly benefit farmers, herbalists, and spiritual seekers.

This and other seemingly paradoxical parables told by Jesus twenty centuries ago still present possibilities that most of us never realize upon our first hearing of his good news.

As we've seen, the nightly gatherings of *haflat samar* in Galilee were the most common means by which the teachings of an elder were imparted to younger women and men in the village, keeping alive the traditions of the community. Peasants could retain the stories, keeping them in their heads and hearts while they went about their daily chores, and then use them as touchstones whenever they faced the daunting challenges of surviving in a hot, dry land.

Take the "Parable of the Tares among the Wheat," one of the most famous of these community stories in the New Testament. The English word *tares* is loaned from a historic Arabic term that means "that which is rejected." It refers to the wild darnel of the Mediterranean basin.

In 1885, a Scottish physician and botanist named John Hutton Balfour first laid out the rich cultural and natural history of this biblical plant. At the same time, he noted that what was remarkable—and also dangerous—about darnel was its ability to mimic the appearance of wheat. The darnel wasn't merely an annoyance. Because it had narcotic qualities and was "noxious," or capable of making people sick, it was important for farmers to be able to distinguish it from the wheat crop.

Given these agrarian points of reference, let's look at the parable of the tares with fresh eyes:

Our Father's kin-dom is like that of a farmer,
one who sowed good wheat seed into his field.
While his household was sleeping,
    the farmer's rivals crept into his fields,
        sowing the troublesome darnel
            in amidst the wheat seeds
before they stole away into the night.

After the wheat had sprouted,
    the farmworkers found it surrounded
        by seedlings of the wild darnel
    with similar leaves of grass.
At first, it sprouted beside the wheat almost unnoticed,
    but even though it mimicked their stalks,
        it carried with it a fungal infestation,
one perhaps toxic to humans and livestock.

When the workers in the field
    spotted the presence of the fungus,
        they ran up to alert the farmer:

"Master, did you not sow good weed-free seed in the field?
If so, how did fungus-infested darnel grasses
    rise up among the wheat in the field
        you wish us to harvest?"

The farmer shrugged and said,
    "Perhaps some rival did this to spoil my crop."
And so they asked him,
    "Should we go ahead and pull them out?"
The farmer thought patiently for a moment
    before he replied,
"That would be perilous,
    for if you try to rogue the wild darnel out of the crop,
        you might uproot the wheat itself.
    Instead, let them grow together
        until harvest time has come.
Then I'll tell all of you who go to reap the yield,
    'Be discerning!'
First, cull the infested darnel grasses
    to bind to bundles for burning,
then gather and winnow the cereal
    for storing in my granary."

The farmer's response has puzzled gospel readers for
centuries. Why not just weed the darnel grass stalks out of

the wheat field as soon as they appeared? Botanist Balfour was puzzled as well, but now we have a few more clues that might explain the wise farmer's decision. Two very different kinds of fungus can infect the darnel, and which fungus you receive makes all the difference in the world.

The grain of the darnel, if taken from plants infested with the fungus named ergot, can make you nauseous, giving you the sensations of being drugged and seeing visions. The ergot fungus—once ritually consumed in the Middle East—was the original source of the hallucinogen we call LSD. Peasants in the Middle East have long recognized both the hazards and/or delights posed by darnel grass that is infected with ergot, since it has the potential to alter the behavior of their livestock and their neighbors.

However dangerous they could be, the hallucinatory properties of ergot-infested darnel or wheat were sometimes used in wildly ecstatic celebrations. Herbalist James Duke, author of *Medicinal Plants of the Bible*, says that some religious practitioners in remote reaches of Mount Lebanon would infuse the leaves of darnel grass or soak its seeds in liquids to extract the fungal element that could trigger powerful spiritual visions.

No doubt Galilean farmers recognized that ergot-infested grasses could be good or bad depending on the context in which they were grown or sometimes consumed, accidentally or intentionally.

And yet there was another fungus—a symbiotic, seed-transmitted one—that also infects the wild darnel coming up as weeds in wheat fields. It is cryptic in the sense that it resides entirely within these grass plants. Scientifically known as *Neotyphodium occultans*, it lives like an organelle in the very cells of the darnel itself.

Simply put, it has no life of its own, except in relationship.

Through a rather astonishing set of processes, the fungus is so embedded within the genome of the wild grass that it can hardly be noticed until the darnel goes to seed.

Remember how the farmer instructed his workers: *let them grow together until harvest time has come.* That's because until their seeds ripen, it is hard to tell which plants were wheat and which were wheat-mimicking darnel grass. It is also because grass infected with this second "occult" kind of fungus serves a vital function in protecting the wheat from pests while the crop is maturing.

This second fungus produces toxins called *loline alkaloids* that spread through the infected plants of darnel. They provide the darnel with a chemical defense against all kinds of insect pests and diseases that are otherwise common in wheat fields. As the toxins in the grass knock the pest populations back or repel them from doing much damage in the wheat field, the wheat yield itself is protected by this naturally occurring biological control.

Get that? The second fungus in the darnel grass is a cryptic "pest deterrent" discernible only at harvest time. It protects wheat—the staff of life in the Holy Land—from myriad perils.

Just how can the knowledge of this relationship work on behalf of farmers? By waiting to see whether the darnel's seedheads are infested by the visually obvious ergot or by the more cryptic loline-producing fungus, farmers and their field-workers can discern whether they have plants that are highly toxic to cattle and humans or only toxic to insect pests!

Keeping the wild darnel that carry the toxic alkaloids in a wheat field can actually reduce yield losses. In short, if an observant farmer in Galilee wanted to control pests in his wheat field, he should wait for the signs of which fungus was present in the wheat and darnel grass, then cull out only the ones that were troublesome.

The mutual relationships or symbioses between fungi and darnel defy any dualistic thinking about so-called weeds. Perhaps it was Matthew—not Jesus—who called them weeds in the first place. Perhaps Jesus recognized the yes/and ambivalence of the presence of darnel, a riddle that herbalists and agro-ecologists have only recently unpacked.

While observant peasants in the Middle East may have figured out some elements of this riddle on their own and put it to use, it has taken scientists centuries to fully comprehend all the mind-blowing complexity of creation embedded in this little story. Their work has revealed yet another dimension of this parable that we could scarcely imagine before this century.

And yet science itself does not "explain away" all the other dimensions of this engaging agrarian story. Instead, it humbly reminds us how much of creation we have yet to appreciate fully and deeply.

Being humbled and awestruck by the "tangled web" of wild nature may be the first step that many of us take toward giving more care to creation, whether we call our actions *conservation* (safeguarding the sanctity and integrity of nature) or *restoration* (healing the wounds in the damaged fabric of our earthen household).

If we read the actions and the words of Jesus carefully, he is calling us to do both. We can keep vigilant about protecting nature's most sacred places and about bringing restorative justice to all cultures and creatures that have fallen in harm's way. Whether we are farmers, fishers, or simply consumers, we are called to care for even the most cryptic or minuscule elements of creation in our midst. Even weeds belong to the kin-dom of our Creator.

11

THE WILD EDGES

At times, all we can do is stand amazed by the weeds and other wild things that intermingle with overly domesticated crops and livestock. I'm struck by the affinity and affection that Jesus expressed for all wild things, including weeds and wayward creatures. He proclaimed an empathy with the ne'er-do-wells who often seem to evade control by authorities, bureaucracies, and industries.

Jesus did most of his preaching either in the wilds or at the weedy edges of villages. He frequented poor neighborhoods "at the dark end of the street" and bantered with his followers *outside* of the markets and synagogues.

It seems he preferred to live on the wild margins of society, alongside lakes and streams, in the shade of hedgerows on the field's edge or under the palms in a remote oasis. That is where the lonely go, as well as the disheartened and dispossessed.

Jesus also hung out by the wells in small villages that provided a dipper of water to thirsty passersby, or where they served cheap wine in the reception rooms of wedding feasts.

When John the Baptist gave Jesus his baptism, he did so by dipping him into a wild river in the depths of a desert canyon, not in a basin constructed in a fancy temple inside the walls of Old Jerusalem.

The death of Jesus took place where he was nailed to a dead tree planted on a rocky bluff at the outskirts of the city, not in a palace court or stadium. His burial was in a small, recently excavated cave, not in a massive catacomb or public mausoleum.

Jesus could fully recognize the Creator in creation, in the exquisite wildness of this world that makes it like none other imaginable. He chose followers who seemed more like weeds themselves than like the manicured trees and shrubs in formal gardens. He knew how to sow wild oats and mustard seeds wherever he went—and called his followers to do the same.

The parable of the mustard seed illustrates that each of us has the potential to become a sower of a small handful of humble seeds of change gifted to us by our Creator. But if someone could then gather our collective efforts together block by block, neighborhood by neighborhood, valley by valley, or continent by continent, they could regreen the entire planet before our very eyes.

About two thousand years ago, Jesus told a gathering of Galilean farmers a parable that offered a wild prophecy:

> What on earth can we compare
> to the kin-dom of our Creator?
>   For me, it is like a wild mustard seed,
> which when tossed down upon the barren ground
> looks like one of the tiniest seeds
>   that could ever fall upon the face of this earth.
> And yet when it is sown in fitting conditions,
>   it grows into an herb so large
> that you imagine it might become
> one of the desert's largest shrubs,
> for it puts out branches that reach toward the sky,
> dense enough for birds to come
>   and take shelter from the sun in its embrace!

I have personally witnessed this phenomenon in the driest of deserts, where, in drought years, the mustard plant of the Middle East hardly grows a foot tall. But in wetter years—merely by drawing upon the moisture stored in the soil from a single drenching downpour—the same mustard stock can produce plants that grow to six to ten feet tall over a single winter.

Hold on to this image of a single minuscule seed of a Middle Eastern mustard plant that in less than half a year germinates and grows to heights that dwarf a tall human. In the spring, sparrows and finches forage for the seeds as they fall from the mustard pods and dry on the ground in the shade of the plant canopy.

This is a lovely image, but in the twenty-first century, many of us miss the downright subversive nature of it. In Jesus's day, Jewish or Phoenician farmers would have seen something else, something a bit darker. As Christian historian John Dominic Crossan points out, the message of the story is not just the astonishing growth of a tiny seed into a towering plant. Rather, it's that a mustard seed "tends to take over where it is not wanted, that it tends to get out of control, and that it tends to attract birds to cultivated areas, where they are not particularly desired."

That's what Jesus says the kin-dom is like. It's something that starts small, on the wild edges, but quickly overtakes everything else. Crossan says that most of us would rather have our dealings with God "in small and carefully managed doses" because we want to keep some semblance of control.

We don't like the messiness, volatility, and uncertainty of true wildness.

What Crossan may be suggesting is that the image of a wildly proliferating mustard is inherently *disturbing* to farmers and farmworkers who must deal with it physically.

But it was also disturbing to the powerful elite in Rome and Jerusalem. Many of them dreaded having to deal with the wild, unbridled anarchism fomenting among an irrepressibly inspired peasantry.

The parable tells those who work on farms that there is something terribly disturbing about the "hidden seed bank" in the desert soil where millions of mustard seeds lay in waiting.

When conditions are finally to their favor, they will "rise out of nowhere" to compete with the elite, more domesticated crops. They will attract large enough flocks of birds during harvest time. They will grow and grow on through late spring, when they will become impossible to control. They will confront the status quo.

Each of them felt like a little seed thrown onto uncertain ground, where they could face insurmountable challenges. Yet if conditions were right, the upswell of undaunted spirit and homegrown power surging among thousands of them could possibly generate change that would unsettle the political and cultural landscape in which they lived.

Jesus needed no militia to make those representing the empire feel uneasy. All he needed was a community of kindred spirits who behaved like little wild mustard seeds in the ways that they grew to meet the challenges before them.

Perhaps we all need to become as resilient as those mustard seeds.

FOOD JUSTICE

Centuries before Jesus walked among us, the agrarian prophet Micah warned us of the threats posed to farmers by the politicos and priests of the bureaucracies that were emerging during his time. Micah's ominous warning of ruthless land grabs during his era foreshadows others that have periodically occurred in nearly every country on the face of this earth:

> In the morning light, they accomplish it,
> > for their hands
> > > are godlike.
> They covet fields and seize them
> > and do the same to homesteads:
> > > they appropriate them.

It is difficult for many of us to emotionally deal with the scale of debt and the other threats that farmers are facing today, just as they did in Micah's time.

If it has never occurred to you that simply owning and tending a family farm might be a social justice issue, a racial issue, or a Christian issue, look again—it is really all three simultaneously. And at the heart of food justice are the stories of some extraordinary people.

12

HERDERS

I n the towns and places Jesus wandered, rested, healed, prayed, and ate, sheep were abundantly present, both in the rangelands surrounding most villages and in the stories the elders told each night in those villages.

Those elders often used their *midrash* commentaries to guide the life and times of ancient Hebrew herders. Sheep run through those commentaries like shooting stars in the night sky.

Jesus would hear the *baaaas* of the enormous flocks of sheep coming down from the hills to the village, where the meat would later feed the feast-goers and their wool would clothe the people. As Jesus sauntered to and fro across the stony, arid, and semiarid lands of the Levant, he often walked amid flocks of sheep kept up in the mountains in summer, then down close to valley wetlands in winter. The roads he walked all began as trails for sheep.

He knew the story of Job, who herded fourteen thousand fat-tailed sheep before his life and fortune fell into ruin.

He fathomed the slaughters that occurred during the weeks-long celebration that marked the dedication of the First Temple in Jerusalem, during which Solomon had 120,000 sheep slaughtered to feed his guests. And he saw and smelled in the Second Temple of his own day how blood was spilled from the necks of rams, ewes, and lambs for ritual sacrifices.

And yet in many of Jesus's stories, metaphorically, "sheep are us." We are the trembling, nuzzling, distractible, fearful creatures adept at the "herd mentality" that Jesus speaks out about. We can sometimes sense that Jesus had conceded that tending "our flock" of sheep was the toughest work that his Father could have given him!

That Jesus chose to take on the humbling, often frustrating, job of shepherding humankind is deeply moving to me. While sheepherding was a common occupation during his era, it was among the lowliest. It remains an underappreciated occupation to this day and one that is now imperiled in many American landscapes.

In John 10, Jesus tries to comfort and calm those who have recently lost much of "the stock" they have worked for all their lives. Whether it has been lost to the predatory jackals and hyenas on the range or the predatory scoundrels in the city, he does not specify. Still, he senses their losses as he walks among the sheepherders, patting their backs and shaking their hands:

I must speak to you as honestly as I can
        about why I have come to live among you:
    I am here to be your gateway into safety,
    like a shepherd who stands by the gate
    to guard the entryway into the sheep's fold
    so that no more robbers and thieves can molest you
        as they have crept in to do
            when you could not hear or see them coming.

    Do you see? I will stand by you
to assure your safe passage, for I am your gateway
    into the kin-dom of heaven-here-on-earth
so that any of you who come with me
    will be nourished by the saving graces
        of this brief life we share here together,
so that you can tranquilly move on to the next pasture.

Do you see?

    I have come here

so that you can live

    more abundantly

        wherever you choose to go. . . .

Jesus identified himself as a shepherd, as if he were nothing more—and nothing less—than the most poorly paid nomadic peasant in the Empire's workforce. He seemed to keep a special place in his heart for every lamb, ewe, and ram—no matter how meek. In the stories Jesus told to his followers, he spoke soothingly to those human "sheep" who felt fragile, fearful, miserable, or vulnerable.

His attentiveness to each individual is what I love most in the stories of Jesus's interactions with his disciples. He kids each one with affection, not derision. He faults himself when he does not reach Lazarus in time before his dear but sickly friend's life seemed to blink out.

You might presume that the Roman Empire had better things to do than to harass a single individual who offered succor and prayers to their society's lowest rung of rural workers. Yet Jesus was considered a threat to the Roman Empire's military and religious leaders for this very reason, for feeling the humanity of the poor. He tried to recognize the unique wounds and gifts of each individual person he encountered. Labels—like Samaritan or Jew, sex slave or tax collector—meant nothing to him. He could care less about pecking orders and social hierarchies. He treated each soul as worthy of respect.

A good shepherd ought to do no less than this.

Two thousand years after Jesus, shepherds and other nomadic herders remain among the poorest, most misunderstood, and least respected professionals in our food system. Herders of sheep, goats, llamas, and other stock number between one hundred and two hundred million people. The higher (and more likely) of these two estimates is roughly equal to all of Brazil's citizenry.

We may not see them every day, but despite the odds against them, *they have not gone away.*

Herders are as diverse as their stock. Basque sheepherders still guide their sheep along trails running from one end of the Iberian Peninsula to the other. In Africa, Berber, Tuareg, and Fulani, herders guide their flocks to patches of grass at the edges of the Sahara, moving hundreds of miles with them each season.

Most herders in America today are hired in from other countries, put out with the stock on remote reaches of summer pasture in western mountain ranges, where they live in tents, line shacks, or Airstream trailers at the end of the trail. They eat canned meats, stews, refried beans, and pancakes and tortillas cooked on a grill. They drink lots and lots of mugs of cheap coffee.

They also get wounded far more often than the rest of us. Broken ribs from being gored by a ram's horn. A bad knee from getting kicked by a camel or pack mule. An amputated foot from being bitten by a rattlesnake. A gashed forearm from being pushed or tossed against a barbed-wire fence. So many herders limp along that one might wonder whether it is in their job description.

They live or die *donde no hay doctor*—"where there is no doctor."

On an individual level, they are physically vulnerable working at their high-risk jobs, but their industry is vulnerable as well. There were once more than 250 million acres of rangelands in the western US, but about 350,000 acres are annually lost from use by livestock as they are converted to residential development, recreational areas, mines, or sites of coalbed methane extraction.

Over the last quarter century, slightly more than half of all private rangeland in the American West has been lost to development. If you roll that out for a few more decades, nearly all free-range cattle, bison, sheep, or goats will be restricted to grazing lands leased from state and federal agencies, lands managed by Native American cowboys on Indian reservations, or pitiful feedlots where corn and hay are given to confined animals.

More insidious and difficult to track is the fragmentation of western rangelands in ways that keep herders from trailing sheep down from summer pastures in the mountains to desert valleys tens or hundreds of miles to the south.

Sheepherders and cowboys can no longer drive their herds across once-open range, for rising land prices force landowners to sell out to developers. Cattle or sheep bring in far less revenue than country clubs, condominiums, summer cabins, or ski resorts.

Some venture capitalists and land developers take livestock producers to be gullible hayseeds, whose land is theirs for the taking. The livestock producers have become underrepresented in political decision-making and not well organized as a group—at least, not organized enough to keep their lands out of the hands of developers who will take those acres out of food production for good.

Other food-producing land is likely to be wrested from local control when speculators buy up ranches and farms only to "flip" them. Through such schemes, roughly a third of all the grazeable, saleable property that is left in the US may be put at risk of never having a cow or a hay crop on it again.

That's where those who stand at the Radical Center come in: the ranchers once tempted to sell their lands in times of financial trouble have met open land conservationists in the middle to cooperate with one another. Since 1993, some two dozen collaborative conservation alliances have sprung up in the American West to put private lands under conservation easements so that they may continue as working landscapes where meat, grass, or tree crops are still produced. They have darkly funny names like the Diablo Trust, Malpai Borderlands Group, Quivira Coalition, the Blackfoot Challenge, and the Shoesole Group. Their goal is to preserve open space and, as one mission statement puts it, "reverse the polarization between ranchers and environmentalists."

But they don't just talk about it. They put their land ethic, their earth-flavored gospel, into practice. Collectively, just seven of the twenty-four collaborative conservation alliances have protected over 3.9 million acres of rangeland for ranching livelihoods (with sheep, cattle, or bison), for wildlife, for wildfires, and for wild-running streams. They've taken down strands of barbwire to allow wildlife to pass through. They host festivals out on the land, field days, range science forums, wildfire watches, and dinners of grass-fed beef.

They've broken the vicious cycle of conflict and litigation by fostering collaboration through working hand in hand together on the land. They are the peacemakers who have stopped the range wars. They have offered unto others the same chance to stay on the land as others have offered them.

If there needs to be a contemporary example of the "radical center" sensibility first pioneered by Jesus, it has cropped up in the semiarid steppes of the American West, where unlikely partners have forged a covenant to work together for the care of creation. Through that restorative process, the wounds of formerly divisive politics have begun to heal as well.

13

WOMEN FARMERS

Jesus seemed to sense how any loss of livelihood often comes with a loss of dignity and self-worth, especially to the poor and to women in particular. He had an uncanny knack for knowing when mothers and wives—not just male farmers and fishers—were losing out on something. He often listened to and accompanied women during their times of hardships.

Because many women outlived their husbands and even their sons, quite a number of them became the household's breadwinners. They tenaciously hung on to any money they received as seed harvesters, fishmongers, gardeners, herders, herbalists, and orchard keepers in order to feed the entire household.

Jesus does not merely celebrate the male shepherd who chases one stray sheep into the wilderness. He also praises the hardworking woman who comes back into the household after laboring outside of it to find a coin misplaced or displaced in the chaos of family life. As Luke tells of the incident involving a lost coin, he paints a picture of a poor but frugal woman who knows how to revel in every windfall that comes her way:

> Jesus said to them,
> "How is God's household like this?
> it is like the place of a working woman
>> who had just ten silver coins
>>> left to her name
> with each of these drachmas worth a day of her work,
>> who worried that one of them was lost somewhere
>> within her household.
>
>
> Will she not take up the lamp, sweep the floor,
>> look under the bedding,
>>> and search within every cup and jar
>>>> until she finds that one precious coin?
> Will she not shout with joy then?
> Will she not call together her family, friends, and kin?,
> exclaiming, 'Let's all have a feast,
>> we have something to celebrate!
>>> I thought we would be busted broke, but now I've found
>>>> the thing of value I thought I had lost!'
> So I tell you, let's embrace that joy that all the angels know
>> whenever they redeem one that has gone astray!"

Jesus shows us in this story that he's not just the Good Shepherd. He's also the Good Housekeeper, expressing how to care for a family and make a living when it seems the means to do so have been lost. And in that sense, he is akin to the women farmers we know.

There are more than a million women farmers and ranchers in the US today who bring us our daily nourishment. That's nearly a third of all the farmers in the country. Who would have thunk it when all we've seen on TV is Mr. Green Jeans or Farmer John? Many of these women also manage their households and a complex web of other activities.

In 2012, my home state of Arizona had the highest proportion of women farmers of any state in our fractured union. But we did not know that until the 2007 farm census, when a lawsuit by tribes forced Arizona to count "more than one farm per Indian reservation," as they had done for decades.

Since that time, census takers have been mandated to do interviews in *native* languages. This meant that for the first time, they counted Navajo grandmothers herding their sheep; Apache cowgirls branding their calves; Hopi mothers tending ancient terraces filled with amaranths, beans, and squash; or Tohono O'odham women growing their own devil's claw plants for basketry materials.

All of a sudden, it seemed like there was a fourfold growth rate in the number of women farming, ranching, or herding in Arizona! Between 2002 and 2007, the state went from having 1,500 farms with women officially listed as the primary operators to over 6,000.

But those food-producing women had been the primary operators of their farms all along. Our government had simply overlooked them, failing to offer them the recognition and technical assistance they deserved.

Today, we can be proud that more than twelve thousand women farmers are recognized as essential workers in our state. At the same time, how sad it is that the majority of these women had been growing crops or raising stock for decades without ever being entered into the government's ledgers. This pattern—once common in many states—is an egregious historic example of structural racism and gender discrimination. A congressional report determined that the US Department of Agriculture has been the last of all the federal agencies to overcome its checkered history of institutional racism and sexism.

Fortunately, other women farmers and herders are being recruited all across the country. My state alone added another 2,429 women food producers to its rosters between

2007 and 2017. These women are of every age, race, and culture, and they are changing the face of American farming and ranching. I have been proud to know some of them and humbled by their consummate skills in producing food and beauty:

- Maya Dailey, an urban farmer in the Phoenix metro area who has at times rented as many as three different fields within the city limits to keep her vegetable supply up with the demand from chefs and other customers;

- Colleen Biakeddy, a Diné sheepherder from Hardrock who direct-markets wool and mutton from her flock of Navajo-Churro sheep;

- Rebecca Routson, a nutritionist and homesteader in the Prescott area who has produced everything from zucchini to aged cheeses to sausages while supporting her daughter Rafael in her efforts to start an orchard of heirloom apples and pomegranates;

- Kate Tirion, a Welsh-born farmer and agro-ecological educator who directs the Women Grow Food initiative at Deep Dirt Farm while producing

rich humus, fine vegetables, and tree crops in
Patagonia, Arizona; and

- Aishah Lurry, an African American farmer of
  gorgeous cut flowers who gardened for nearly two
  decades before starting the Patagonia Flower Farm.

Along with the rise in women farmers in my home state,
we're witnessing a 27 percent increase in the number of
women working as farm operators across the nation. From
2012 to 2017, more than a quarter million women joined the
ranks of US food producers. More than half of all American
farms now have a woman as sole or co-manager. Their farms
account for 38 percent of agricultural sales harvested off
of 43 percent of US farmland. Nevertheless, they still face
daunting challenges.

Carefully listen to the words of Jesus for what they say
both about women themselves and about their importance
to the integrity of farming families:

I swear to you,
there should be no one among you
    who has left your brothers and sisters,
        Your mothers or fathers,
    who has forfeited your farm

on account of your bonds with me
or for the sake of the good news I bring you
who will not receive a hundredfold greater harvest
    later on for the keeping of your faith.

In this age, you may suffer
    through persecutions and dispossessions
    of the kinds that threaten
        your herds or homesteads.

Yet in the kin-dom to come,
    there will be life everlasting
        so that many who are now first
            will be put last
            while the last
        shall be put first.

Whenever I see Colleen Biakeddy with her sheep or Aishah Lurry with her cut flowers, I feel hopeful that more and more minority farmers are being celebrated and not left on the margins to fend for themselves. These are strong, intelligent women who integrate skills, talents, and values I seldom see expressed as fully in male farmers or in urban gardeners of any gender. They live and breathe with a sensibility that land is sacred and that caring for the land is tantamount to a spiritual calling. My Hopi friend Michael Kotuwa Johnson calls this sensibility "faith-based farming" because it takes conviction to do such work in the face of all the uncertainties and challenges that farmers face today. As women farmers bless the land with their diligent work, they are in turn blessed by its bounty, in a reciprocal but silent prayer that makes this earth a better place to live.

14

GLEANERS

I t is likely you know someone—though you may only rarely recall her or his name—who gleans and eats from the leftovers of others.

The Qur'an claims that Mary, the mother of Jesus, worked as a gleaner in the floodplain fields along the Nile after her family was forced to flee from the Romans into exile. It is likely that most of the gleaners during her time in Egypt were young women.

It is said that she carried the baby Jesus as she gleaned the unharvested grain. There's an apocryphal story that claims that when she dropped him down on the edge of the fields, the infant Jesus played in the mud along the irrigation ditch, shaping doves out of clay. They flew away!

Gleaners today are everywhere but are often invisible to others—or worse, harshly judged as opportunistic scavengers. Perhaps we need a better way today to remind ourselves that the act of gleaning is never as demeaning as being hungry is or as being prevented from accessing fresh food. The larger tragedy in America today is that roughly one hundred billion pounds of food each year is wasted or lost between the field and the table. That's nearly a third of all the foodstuffs produced on American soil.

At the onset of the COVID-19 pandemic, when one out of six Americans was thrown out of work without easy access to affordable food, tons of perfectly edible tomatoes, peppers, and other fresh produce were being dumped in landfills or left in fields. If those foodstuffs could be fully rescued and distributed, they would be sufficient to meet the nutritional needs of nearly every hungry person in our country. And yet, whether we can cost-effectively mobilize the rescue of fresh foods in a timely manner on a massive scale remains a big *if*. As I spent time at a food bank bagging tomatoes for curbside pickup during the coronavirus lockdown, I realized that our communities each need hundreds of more volunteers in order to redirect food away from dumps and into hungry bellies.

Gleaners get used to waiting. In first-century Egypt, where Mary and Joseph were living as refugees when Jesus was very young, there were customs in place to regulate the practice of gleaning. They could only go into the fields for the remains of the food crops once they had been officially "sanctioned" as waste. They had to wait until after the scribes and religious authorities confirmed that the Pharaoh himself had fully received his mandated share before they could glean from what had been left unharvested.

As one account describes, then and only then could long lines of reapers begin to move in unison down the planted rows: women, children, old men. Artwork found inside an Egyptian tomb depicts a woman gleaner holding out a hand and pleading, "Give me just a handful. I came last evening. Don't make my luck as bad as it was yesterday."

Perhaps Jesus witnessed such a scene while straddling Mary's hip or when resting his feet in the cool mud as he sat on the edge of a ditch. Perhaps it was somehow burned into his memory:

My own blessed mother, Mary, had to glean
    the food that other people threw away
        just so we could eat.

Jesus may have remembered such moments three decades later when he and his friends were out on the road around the time of the harvest of cereals and beans in late spring or early summer. It was already hot, and they were hungry.

But it was also the Sabbath, and they had been brought up with the notion that this was a day when all work was forbidden, even if it was required to keep themselves alive.

And it so happened
　　that as he was walking along
through fields of grain
　　on the Sabbath day,
　　　　his gang of famished disciples
　　　　　　began to strip the clusters
of seeds off stalks of crops as they hiked along.
　　　　　　When the Pharisees heard this,
they confronted Jesus over it, shouting,
"See here, why is your mob doing exactly what
has never ever been allowed by Law on a Sabbath day?"

But Jesus responded,
"Haven't you read the text
　　of what David himself sanctioned
　　　　when he found that his crew had empty bellies?
　　David walked right into the House of the Lord [at Nob],
　　the one where Abiathar would soon serve,
　　　　and David snatched the consecrated bread
　　　　　　right off the altar
　　　　　　　　to give to his friends to eat.

Is no hungry soul
  ever permitted to taste
    the bread of this kin-dom
      unless he's born into a clan of high priests?"

A number of Christian historians and theologians have linked this gleaning event to other events later in Jesus's life when he opted to meet people's fundamental needs, even if that meant doing so by breaking the rules.

Jesus would even consider suspending a tenet of the Sabbath if the rules for its observance limited poor people's access to food and resources. He refused to use the Sabbath as a weapon to deny the most marginalized people a place at the table. He loved, and looked out for, the lowliest of gleaners.

Most farmers are pained when the perfectly edible food they grow has to be left in their fields. They are sometimes economically forced to plow under crops produced from their own labors so that the food can never be accessed, even by gleaners working for food banks.

Still, there is another more insidious kind of food waste that happens in the home. Your home. My home.

Sadly, a seventh of all the food purchased by consumers in Tucson (where I lived and worked for decades) is tossed out within two weeks after the grocery bags are brought into the household.

The same trend occurs at the national level: the average American family annually tosses out about $600 worth of once-edible vegetables, grains, fruits, and meats after bringing them home from the grocery store, farmers' market, food bank, restaurant, or garden.

Fortunately, Tucson is also home to one of the country's most efficient gleaning projects. The Iskashitaa Refugee Network alone salvages tens of thousands of pounds per year of edible fruits from the backyards of Tucson. Recently arrived immigrants involved in the network then transform that food into delicious, value-added products.

Most of this transformational work is done by refugees who are still trying to fathom just why there is so much waste in the US food economy. The majority of Iskashitaa's workforce is women—including ones who were forced to leave Syria, Sudan, Somalia, or other drought- and war-stricken countries in dire straits.

After years of facing hunger, these immigrants are not as inclined to allow food to go to waste as their better-settled neighbors have been. And so they glean and then process tons of fruit into sauces, vinegars, juices, and syrups for sale in their community. Perhaps they are here to shame us into realizing how much we waste as others hunger.

The Iskashitaa Refugee Network and other food projects in Arizona dramatically reduce the magnitude of waste in our food system. But they also shine a light on the many hidden faces in our food system, showing their tremendous diversity.

The racial diversity as well as the very humanity of gleaners in my state have been immeasurably enriched by recent refugees and immigrants. They have come from Afghanistan, Bhutan, Burma, Burundi, Central African Republic, Cuba, the Democratic Republic of Congo, El Salvador, Equatorial Guinea, Egypt, Eritrea, Ethiopia, Guatemala, Honduras, India, Iran, Iraq, Japan, Liberia, Republic of Congo, Russia, Somalia, South Sudan, the Sudan, Syria, Togo, and Uzbekistan. The same trend is evident in many other states as well.

Remarkably, the national Feeding America network now involves individuals from dozens of cultures who rescue 3.6 billion pounds of still-edible foodstuffs from fields, orchards, vineyards, and restaurants in America each year.

That's almost a third of the amount of otherwise-unharvested food left on the ground and unsold in our agrarian landscapes. Such rescued food goes to soup kitchens, church pantries, food banks, and Meals on Wheels programs that serve the poor, the elderly, the disabled, and the incarcerated.

Supporting food banks with gleaning programs is one place where Christian, Jewish, Buddhist, and Muslim communities truly step up to the plate, hardly ever asking whether the recipients of this food are from their own faiths. Few of the donors are even concerned whether the food ends up in the mouths of still-undocumented workers or in the mouths of those who have achieved full citizenship.

It is easy for some Americans to forget that many of the gleaners themselves are hungry or homeless. *These are no slackers.* Of 110,000 pounds of orchard fruits and field produce harvested in one county by volunteer gleaners, 77 percent of the harvest was donated to local emergency food relief programs, but another 23 percent was taken home by gleaners to eat themselves, to share, or to process for later use.

Fortunately, there have always been farmers and fishers of all faiths who simply offer their unsold harvests to the poor.

During the height of the Great Recession of 2008, one farmer near Denver opened his entire potato farm to gleaning by whoever came to do the work, no questions asked. To his surprise, twenty thousand people arrived over the following days, many expressing their gratitude for his generosity.

Gleaning happens in our waters too, not just on land. In Vancouver, a Native American fisher offered five hundred sockeye salmon to the hungry in an encampment for the homeless.

Perhaps those most directly involved in fishing, hunting, and harvesting the many wild foods that bless our continent are the very people who are least likely to allow that bounty to be wasted. In nearly every village, town, and city in North America, you can find such people of modest means who reliably and unerringly donate any "windfall" foods they come upon to those who are less fortunate.

These people take time to act in the very manner that they attribute to David, Jesus, or the prophet Muhammad, leaving no stone unturned in efforts to nourish and heal the hungry around them. Ensuring that the hungry are fed is perhaps the oldest and simplest way that a member of our species can express the compassion that is at the root of our faith.

Assuring that others are nourished is what we are called to do, just as much as the mother bird is called to bring her open-mouthed chicks a cricket or corn earworm back to their nest.

15

PROFITEERS

In the Second Temple in Jesus's day, an unimaginable number of pigeons were caged and sold for sacrifice as holy offerings. These domesticated doves were not consumed as broiled squab for all to share but instead were burned as offerings demanded by the priestly class.

Not far from the cages of pigeons on an ordinary day at the temple two thousand years ago, you might witness dozens of money changers converting the coins of the Roman Empire into Jewish currency. That was the only kind of currency that could pass from the outer marketplace into the inner sanctum reserved for worshipping by the Jewish elite.

There, poor widows were forced to pay for an unfledged pigeon (or "squab") for a sacrifice, for a tithing to a priest, and even for physical entry into the second tier of the temple. All Jews were obligated to make sacrifices at the temple, and the burden for doing so hit the poor the hardest.

You can bet that Jesus was wary of what tended to happen whenever organized religion combined with the extractive economy. He loved his faith and its sacred teachings, but he shook with anger when those were obscured by market intrusions into the temple.

This story about Jesus is present in all four gospels, which gives a sense of its importance to the disciples who knew him. The exploitation that happened in the temple was enough to incur wrath from Jesus, the Man of Peace. Here I offer a harmonization of the different versions of the story, describing Jesus's condemnation of the unholy union between religious institutions and economic empires.

> When he came into the temple, Jesus became enraged:
>> he began driving out all vendors,
>>> shoppers, and money changers
>>>> out from the outer reaches of sanctuary,
>> for he had seen enough of those who were selling
>>> lambs and squabs on behalf of their bosses.
>>>> He could no longer stand all the money changers
>>>>> doing their business as usual for the empire.

Witnessing all this, Jesus hastily made a whip out of cords
        and began to disrupt everything going on before him,
overturning every cart, table, seat, and cage.
        He spilled their piles of coins onto the floor,
        forcing the money changers to scramble on their knees
as they vainly tried to snatch up a few coins
        before fleeing to escape his wrath.

He did the same with the squab salesmen
        who had been selling squab to widows, shouting
        "Take this junk away, and release those squabs!
You have no right to murder my Father's creatures—
        nor transform my Father's house—
        a Sanctuary of the Sacred—
        into a secular marketplace!"

Jesus quoted the prophets Isaiah and Jeremiah,
        shaming them with their own holy texts.
These texts that made it clear that the temple
        was meant to exclusively serve as a house of prayer
for all people. Instead, it had become
        something else entirely:

        "Now it's become a haven and holdout for scoundrels,
        for ruffians who slyly rob the poor
and do violence to the sacred!"

Pharisees stumbled onto the scene, shouting at him for disrupting their commerce and demanding to know what authority he had to do so. *How dare he?*

It's likely that this very conflict with the authorities over profaning the sacred spelled the beginning of the end for Jesus, who would be dead within a week:

> When the chief priests and the scribes got wind of all this,
> > they knew they had to take him out.
> By this time, they were so jealous and envious
> > of how his teachings had astonished so many
> that they feared that the multitudes be aroused
> > and would continue to grow
> until someday, the masses might stand up
> > and take out the elite who profited off the temple.

In his controversial book *Zealot*, Reza Aslan unpacks the conditions that raised Jesus's ire in the temple that day, anger the likes of which we see in virtually no other story in the gospels. Aslan says that when Jesus called the temple a den of thieves, he wasn't actually saying that the people who sold animals for sacrifice were the corrupt ones. Their presence at the temple was customary, and they provided a necessary service. Rather, "He indicted the temple authorities as robbers who collaborated with the robbers at the top of the imperial domination system." Jesus was raging not at the merchants and money changers themselves but at "those who profited most heavily from the temple's commerce, and who did so on the backs of poor Galileans like himself."

This incident led directly to the crucifixion of Jesus. Jesus had gone to the heart of the problem of any corrupt religious institution, indicting those within and those beyond the Judaism of his era. He knew by that time that his last stand would be upon a cross, along with so-called zealots or other enemies of the state.

Ironically, Jesus was not at all advocating for a political uprising to take over the temple or the empire as the Pharisees feared. He was not threatening armed resistance. What Jesus espoused was something far more humble, hopeful, and potentially more lasting: spiritual defiance that was grounded in a sense of contemplative action and restorative justice. He served notice to the profiteers that Yahweh cared deeply about justice and that any prayerful person should not permit the temple to be reduced to a place of exploitation. He moved to restore its sanctity and make such hallowed ground accessible to all people, rich or poor.

LAST SUPPERS

How do you get over feeling brokenhearted when your proverbial crop has failed or your most high-minded calls for justice have fallen on deaf ears? *Such broken hearts are pervasive signs of our troubled times.*

As Jay Parini reminds us in *Jesus, the Human Face of God*, things did not always go well for Jesus during his fleeting moments on this earth. Sometimes he seemed overwhelmed, if not physically repulsed, by the dysfunctionality and disparity he saw around him in the boom towns of Autocratoris and Tiberias or in Tyre, Capernaum, and Jerusalem. Coming down from the mountains after his transfiguration, he descended into one kind of trouble after another, leading quickly to his death at the hands of the Roman authorities.

But this was not the end of the story. Jesus's tragic death was first read as a sign that the ruling powers of this world had claimed yet another victim of injustice, but his resurrection and the subsequent flourishing of communities of the faithful were entirely unexpected. These events revealed the kind of reversals that God made possible. Jesus not only upended the carts in the temple; he upended the uncaring systems that dominate the secular world. His followers have done the same again and again. They offer new hope for every victim of oppression, every farmer and fisher and gleaner who has ever dreamt of a life defined by membership in a caring, redemptive community.

16

BARREN FIGS

After being disappointed by what he perceived as the dry and barren spiritual life that had come to dominate Jerusalem, perhaps Jesus hungered for something nourishing, luscious, and thirst-quenching.

In his world, the fruit of a fig tree could often be counted on for this purpose. But in the very same week that he felt overwhelmed by the spiritual aridity of the temple, he had another disappointing experience in his search for a mouth-watering fig. The two stories are wound tightly together like crisscrossing strands of rawhide in the gospels. Hoping for sustenance and a brief repast on his way into Jerusalem with the other pilgrims for Passover, Jesus spotted a fig tree on the side of the road. Upon first glance, it too appeared to be "barren," lacking any seed-laden fruit.

And yet first appearances can be deceiving, and a "barren" fruit might just have life and hope after all.

If the story about Jesus overturning the money changers' tables in the temple makes evident his capacity to spill over with anger about chronic injustice, his encounter with the fig tree confirms his own vulnerability to despair.

As they were leaving Bethany,

         Jesus got hungry.

He looked around until he spotted a fig tree

      there in the distance, with new leaves on it,

so he went toward it,

      expecting to find some early figs.

But when he got right up to the tree,

      there was no fruit yet in season,

           just the sprouting leaves.

Jesus reacted passionately, cursing,

      "May no one so much as taste

        your fruit ever again."

His disciples took notice

      because that was the first time

        they had ever heard Jesus curse.

Note that just before seeking out the small but sweet fruits of spring, Jesus had gone to the temple to be spiritually comforted or inspired. At the very least, perhaps he would have liked to be consoled by the preparations for what he sensed might be his last Passover. But all that he witnessed in the holy city was "business as usual."

In the heart of the metropolis, Jesus found nothing that impressed him spiritually. As he moved out along the pilgrimage route where fig trees grew on the edges of small orchards, nothing around him seemed to be bearing fruit or producing seeds that could sprout into a more hopeful future.

Both the temple and the fig tree seemed barren, as if they were altogether lacking in something fundamentally nourishing. But I bet that his disciples would be delighted—as I have been—by learning the botanical context of spring-bearing figs. When we know some of the botanical background, we can see that a so-called barren tree may have vitality left in it.

See if you can focus in a fresh way for a moment on the peculiarity of the barren fig tree as a metaphor, especially as one meant to reflect the barren temple and its impoverished faith community.

As Jewish farmers and gardeners of Jesus's era readily understood, only a few varieties of figs flower and leaf out in the early spring season around the time of Passover. One particular kind might even bear fruit by the time of the Passover, and they are quite unlike your typical figs.

This peculiar variety is known horticulturally as a *breva* fig, for it offers a very brief season of harvestable fruits before any other kind of cultivated tree begins to bear. It produces unusual fruit in the spring on older, harder brown branches.

In fact, the breva type cannot produce fruit on the fresh new green wood of the tree, where the larger, sweeter figs of late summer and autumn will appear.

These knobby but still flavorful figs have long been known in the Semitic languages of the Levant as *taqsh*. That is a term still employed by Arabic-speaking Jews, Muslims, Druse, and Christians in Galilee today. Taqsh are produced only on a small group of breva fig varieties, including the San Pedro (Saint Peter) fig and the broadly cultivated Black Mission fig, but both were widely grown in the Mediterranean basin during Jesus's time. Peasants would eat the taqsh when hungry, enjoying these knobby snacks during the brief season of spring.

The taqsh themselves are sometimes thought of as "barren" figs because they bear neither seeds nor any viable progeny. They develop on the fig tree, not as a result of sexual reproduction, but rather by the rarer asexual process of *parthenocarpy*. That means a fruit can fully mature even though it is seedless, without the benefit of fertilization of its ovules through sexual reproduction.

Yet parthenocarpy does not actually imply that the fig trees are barren. Instead, orchard keepers often describe them as having "virgin fruit." They mature through means other than the normal joining of genes from male and female floral organs. What an interesting "actor" to briefly appear on stage in the era of a prophet who was believed to be the product of a virgin birth!

The resulting fruits on breva fig trees in the spring are therefore seedless but fleshy enough to be edible as a famine food. They are often relished because the fig trees yield well at a time of year when few other fresh fruits are available to nourish the hungry. In the Middle East, many fig lovers wait in anticipation for these "first fruits" of the season. They have less sweetness but still offer a substantive flavor in their flesh.

It may be no coincidence that Jesus sought out these fruits to make a point.

Did he feel something for these figs that were present on the margins of the metro areas of the empire? They were found less than eight miles from Jericho and just twenty-five miles away from Bethany and Jerusalem. Even today, such "virgin" fig trees still cling to canyon walls above the trails around Jericho and the Jordan River, nourishing the hungry who ascend to caves and monasteries there to meditate.

Jewish peasants of his era may have assumed that the lack of seeds was due to neglect. Or perhaps the fig tree Jesus observed had already been stripped of all its taqsh. Maybe some figs had been plucked and plopped into the mouths of hungry pilgrims or sold into the markets near the temple just before Passover.

What a remarkably poignant moment in the life of Jesus! He too may have worried whether his own life had been barren, that his teachings and friendships were for naught. Perhaps he despaired whether his all-too-brief life on earth had been as fruitless as the roadside tree or as the listless community that had begun to gather for Passover around the temple.

Perhaps he too felt "out of season," as if his life and ministry would not contribute to lasting change within the political environment he had been born into.

No wonder Jesus cursed the fig. And no wonder we can palpably feel his disappointment and anger in the next episode of the story, when he enters Jerusalem and rages at the extractive economy that used the temple as its hub.

But let's stay here with the fig tree. Orchard keepers know that even an apparently barren fig tree that is well stewarded can suddenly produce fruit, even out of season. This kind of resurrection or rebirth is not some far-fetched abstraction but a tangible reality that you yourself can taste.

Here's what Jesus himself said about resurrecting apparently barren fig trees in another parable:

A man had a fig tree
    growing in his vineyard;

he came looking for fruit on it,
    but he didn't find any.

So he said to the vineyard keeper,
    "See here, for three years in a row
I have come looking for figs on this tree,
    but I haven't found any.
Cut it down, for why should it suck
    nutrients out of the earth
        other plants can better use?"

In response, the vineyard keeper said to the man,
    "Let it stand for one more year,
        until I dig around the base of its trunk
to work in some manure.
Maybe it will produce next year,
    but if it doesn't, well, let us go ahead and cut it down."

Jesus did not offer to tell his disciples what the outcome would be for the figs on that tree, nor for their shared kindom on earth. Instead, he offered them the hope that diligent care of the tree of life—of the ever-branching families of humans and other species around them—might bring to fruition something beyond our wildest dreams.

17

COMMUNION

For two millennia, Christians have faithfully practiced the ritual of the open table—one where everyone is welcomed into a place where they can be nourished. It is a rite that dismisses the conventional boundaries that protect exclusivity and generate disparity whenever economic powers decide who should be invited in for a warm meal and who should be left out in the cold.

Whether we recognize it, each time we share Holy Communion with our friends, families, and neighbors, we engage in a rite that is at the same time revolutionary and reconciliatory.

In solidarity with the hundreds of millions of other souls who have dared to break bread together and sup from the same cup in defiance of the forces that seek to divide us, we still share a meal.

Communion was not meant to be an exclusionary practice restricted to bona fide members of a single community, intended to benefit only the in-crowd. It encourages us to join hands with all of humanity—and paws with all of creation—no matter what other expressions of faith may be among us.

The contemporary practitioners of this Eucharistic rite might be refugee farmers ensconced deep within a basement hideaway in Aleppo, Syria, wondering just whose bombs and missiles are rattling their neighborhood at that particular moment in history.

They might be undocumented farmworkers who have briskly walked in to worship in a rural church, knowing that law enforcers from the Immigration and Naturalization Service will be waiting in their trucks just outside the sanctuary when they exit into the harsh light of day.

They might look like a couple of gay sous chefs suddenly dismissed from their jobs at a restaurant when its manager decides that what he perceives as their conspicuous presence might not be good for business in a small conservative town.

Whenever we all come together for rituals such as the breaking of the bread and sharing of the cup, each of us in our own ways, in our various stations all around the earth, are vibrantly responding to and reverberating this simple request from Jesus:

Do this in remembrance of me.

What we call the "Last Supper" was the final meal Jesus shared with his disciples before his crucifixion. This is the scene depicted in famous works of art and memorialized for the last two thousand years by Christians who reenact his sharing of the bread and wine as their Eucharistic rite of Holy Communion.

But there was actually another "final" meal in the New Testament, a "later supper" that Jesus shared with his friends on the shores of the Sea of Galilee. It occurred sometime after his crucifixion and resurrection. His friends had gone fishing that night, but like that first time they met Jesus, they hadn't caught a thing.

As the first hints of dawn reached the weary fishers, they spotted the silhouette of a figure on the beach as they rowed their boats into shore. That is when the drama begins:

> They heard a voice calling out to them,
> but they could not make out whose voice it was:
>> "Hey fellas, you haven't caught a fish
>> you can share with me, have you?"
> "No luck, no catch," the one closest to him replied.
> Standing on the shore, Jesus laughed
>> as if he had heard that one before,

for they had said the same that first day

    he had made them Fishers of Men.

And yet he urged them to go back out, pointing the way:

    "Look here: cast your nets off the right side of the boat.

Yes, right there! And see if you'll have any better luck."

    Reluctantly, they did what he suggested, and instantly,

they felt so much weight in the net

    that they could hardly haul it up.

        It was loaded with a teeming abundance of fish!

One of them shouted out to Simon Peter,

    "Good lord, it's the master! It is the master!"

When Peter heard these words, he tied his cloak to his torso

    and jumped into the water, for he had been buck naked.

As he guided the burgeoning nets in toward the shore,

    the rest of the crew came in the boat, rowing so as

to drag the nets into the shallows without ripping them.

    When they arrived at the beach to secure the catch,

they noticed a wood fire burning brightly there,

    where someone was preparing

to reheat the bread and grill the fish.

    Jesus beckoned to them,

"Bring me some fish from your catch,"

    which Peter did, hauling the nets out of the boat.

And counting some 153 fish,

    Peter brought them onto the landing.

Even though they were exceedingly heavy,

    the nets did not tear.

As Jesus turned to them,

    they could feel his exuberance,

for he was smiling as he beckoned them,

    "Come, let me feed you!"

Despite the doubts and the uncertainties that lay before them, Jesus's disciples sooner or later were reassured that there was still an abundance in the world that could be fully accessible to them.

Like the first time Jesus had instructed Peter to direct his boat over to where fish clustered near a hidden set of springs, they were surprised by just how bounteous their home place could still be. But Jesus had to constantly remind them that they were cared for, not abandoned, and that he would look out for their needs.

It was through Jesus's gracious preparation of bread and fish—not bread and wine this time—that his disciples realized they were capable of living and participating in the presence of unanticipated miracles, sudden windfalls of abundance, and joyous feasts of sustenance. They knew they should nourish not just their own bellies and hearts but also those of thousands of others who groaned with hunger and despair.

His disciples were so delighted to celebrate with him again that they may have forgotten to bring out the wine. And so that "resurrected" communion may have simply consisted of bread and fish.

Bread, wine, fish. Sacred soil, the blessed ferment of spirits, and the holy sea.

Perhaps that dynamic triad of bread, wine, and fish can become for us another kind of Holy Trinity, one that brings together the gritty earthiness of wheat bread, the heavenly ferment of grape wine, and the oiliness of sea creatures into a moving sacrament that changes before our eyes.

In this way, we can be reminded that our Creator so loved the world that we are offered a full communion with every creature and culture on the land, in air, or afloat among the seas.

Jesus broke the rules of exclusion and purity that separated Pharisee and scribe from peasant farmer, fisher, farmworker, and slave.

And that's why, not long after his reappearance among his fishing and farming friends, he invited all of them to share in the wild abundance of creation as it morphs and buds out, flowers, flourishes, and fruits to regenerate itself once more! Creation is never over, never! Every farmer and fisher of every race and age lives and labors in the midst of it, lending their individual and collective hands, hearts, and dizzy heads to that divine creative process.

The community of disciples blessed by Jesus's example continues to grow with the astonishing fecundity of little mustard seeds. You are part of that flock. You can lift your voice to be part of that breathtaking choir of all creatures. As we sing grace together, more bread, wine, and fish just keep appearing on the table, nurturing an ever-greater diversity of peoples who share in them. The foods of this blessed earth and its sacred seas are the precious, delicious gifts that most directly connect us with creation and with the wild love of our Creator.

ACKNOWLEDGMENTS

I am spiritually indebted to my prayerful, playful, and joyfully musical wife, Laura Smith Monti, as well as my brothers and sisters in the Order of Ecumenical Franciscans. They have been my anchors during the somewhat tumultuous journey I have been on the last six years.

Each of my all-too-brief but fruitful encounters with impassioned spiritual directors over the decades—Father Richard Purcell, Father Dave Denny, Mother Tessa Bielecki, Nelson Foster Roshi, Sister Nancy Menning, Sister Dale Carmen, Brother Carlo Gendron, Brother Michael Vosler, Brother Juniper Robertson, Amadeo Rea, and Douglas Christie—has offered guidance and encouragement of a nature that has surpassed anything expected in a purely professional relationship. Blessings to you all.

This narrative emerged out of life changes gifted to me by Brother Mule! It led to my enrollment as a student in the Living School associated with the Center for Action and Contemplation. I am so grateful to Father Richard Rohr, Cynthia Bourgeault, Barbara Holmes, and James Finley for their fine contemplative teachings and goodwill, to their support staff, and to my fellow students for their camaraderie. I particularly wish to give a nod to my friends Tom Eberle, Susan Lamb Bean, Mandy Montañez, and Rhett Engelking for being among the first readers of my earlier drafts. I have also benefited from two summers at the Glen Workshop West for writers "of spirit" at St. John's College in Santa Fe, where I was enlightened by brief but incisive conversations with Scott Cairns, Richard Rohr, Deborah Smith Douglas, and Scott Russell Sanders.

My conversations over the years with several humble visionaries have consoled and convinced me that the nexus between spirituality and food justice is where I need to be the rest of my life. These people include Fred Bahnson; Victoria Loorz; Anna Woofenden; Wendell, Tanya, and Mary Berry; Douglas Christie; Jennifer Abe; Curt Meine; Norman Wirzba; Michael McDonald; Veronica Kyle; Jim Mason; Jose Oliva; Mary Evelyn Tucker; Abbie Rosner; Joel Salatin; Kristy Nabhan-Warren; Robin Kimmerer; and Beorge Ballis. I must thank Brother Fred for helping place this manuscript in the hands of Lil Copan, a gifted editor who immediately got what I was trying to do, who then pledged to help me get it to where others could make sense of it as well. Special thanks to Jana Riess, Lil Copan, Claire Vanden Branden, Marissa Wold Uhrina, and Mandy Montañez for helping me see where the metaphoric holes were in the wool sweater of this manuscript. Jana did an especially thorough and thoughtful job of helping me get both historic content and syntax right, well beyond what I could have possibly achieved on my own.

As I have matured with age, I have realized that strictly secular, materialistic solutions to such complex, pressing, and pervasive issues as those facing our food security just don't seem to cut it. I remain grateful for what I have learned from these friends and mentors and for their own contributions to what my Hopi friend Michael Kotuwa Johnson calls "faith-based farming."

NOTES

Introduction

5 **Their destitution had become crippling:** Douglas E. Oakman, *Jesus and the Peasants* (Eugene, OR: Cascade Books, 2008).

6 **Death of Herod the Great:** John Dominic Crossan, *Jesus: A Revolutionary Biography* (New York: HarperCollins, 1994).

11 **Jesus never mentioned this metropolis:** Marcus J. Borg, *Jesus: Uncovering the Life, Teachings, and Relevance of a Revolutionary* (New York: HarperCollins, 2006).

14 **Jesus had a poet's peculiar capacity:** Richard Rohr, *The Good News According to Luke* (New York: Crossroads, 1997).

Chapter I: What Was Happening *Then* Is Happening *Now*

23 **Agrarian crises from the Bible:** Ellen F. Davis, *Scripture, Culture and Agriculture: An Agrarian Reading of the Bible* (New York: Cambridge University Press, 2009).

23 **The same patterns happening in our time:** Wendell Berry, *The Unsettling of America: Culture and Agriculture* (San Francisco: Sierra Club Books, 1977).

23 **The demise of the family farm:** Victor Davis Hanson, *Field without Dreams* (New York: Free Press, 1996).

24 **"Death on the farm":** Max Kutner, "Death on the Farm," *Newsweek*, April 10, 2014, https://tinyurl.com/yckg772w.

24 **Four-fifths of the world's food:** Chris Arsenault, "Family Farms Produce 80 Percent of World's Food, Speculators Seek Land," Reuters, October 16, 2014, https://tinyurl.com/ybrxy3vn.

25 **Ratio of farm income over expenses:** Claude Hummel, "Impact of Higher Expense Ratios to Agriculture," *Journal of Farm Economics* 36, no. 5 (December 1954): 1002–1008.

26 **Jesus and his disciples did not take up arms:** Bruce Chilton, *Rabbi Jesus: An Intimate Biography* (New York: Random House, 2002).

27 **"You cannot solve the evils of the world":** Eckhart Tolle, "Toward a New Earth: A Conversation with Elizabeth Lesser," Eckhart Tolle Now, accessed May 28, 2015, https://tinyurl.com/zkbsz6g.

Chapter 2: The Sower

39 **Unpredictable outbreaks of crop pests:** T. W. Mew, E. Borromeo, and Billy Hardy, *Exploiting Biodiversity for Sustainable Pest Management* (Los Banos, Philippines: International Rice Research Institute, 2001).

40 **A heterogeneous mix of seeds:** S. Grando, R. Von Bothner, and S. Cecarelli, "Genetic Diversity in Barley: Use of Locally-Adapted Germplasm to Enhance Yield and Yield Stability," in *Broadening the Genetic Base of Crop Production*, ed. H. David Cooper, C. Spillane, and T. Hodgkin (London: CABI, 2001), 350–368.

Chapter 3: Seven Springs

51 **Heptapegon:** David N. Bivin, "The Miraculous Catch: Reflections on the Research of Mendel Nun," *Jerusalem Perspective*, March 1, 1992, revised January 6, 2013, http://jerusalemperspective .com/2644/.

52 **The cords of trammel nets:** David N. Bivin, "Let Down Your Nets," *Jerusalem Perspective*, January 1, 1990, http://jerusalem perspective.com/2451/.

Chapter 4: Sustaining Abundance

57 **The fish populations of James Bay were stable:** Fikret Berkes, "Fishery Resource Use in a Subarctic Indian Community," *Human Ecology* 5 (December 1977): 289–307. See also Berkes, *Sacred Ecology* (New York: Routledge, 2002).

60 **Scholars haven't taken the time to explore Galilean fishing:** K. C. Hanson, "The Galilean Fishing Economy and the Jesus Tradition," *Biblical Theology Bulletin* 27, no. 3 (August 1997): 99–111.

64 **Many native species have gone extinct:** Moshe Gophen, "Extinction of *Daphnia lumholtzi* (Sars) in Lake Kinneret," *Aquaculture* 16 (January 1979): 67–71.

64 **These fish had lost sight in both eyes:** Moshe Lichtman, "An Unknown Virus Is Blinding Fish in the Kinneret (Sea of Galilee), Causing Them to Starve to Death," *Globes: Israel Business Arena*, October 24, 2011, http://www.globes.co.il/ en/article-1000691946.

65 **In 2010, Israel's ministries had to ban all fishing:** Nathan Jeffy, "Fishing Banned on the Sea of Galilee," *Telegraph*, April 3, 2010, https://tinyurl.com/yk9r4ph.

Chapter 5: Changing the Way We Fish

70 **The sea lamprey was accidentally introduced into the Great Lakes:** Laurie Sommers, Fishtown: Leland Michigan's Historic Fishery (Traverse City, MI: Arbutus, 2013).

71 **Four out of every ten species of freshwater fish:** H. J. Jelks et al., "Conservation Status of Imperiled North American Freshwater and Diadromous Fishes," Fisheries 33, no. 8 (August 2008): 372–407.

73 **"The world's waterways call us to practice social justice":** Diana Butler Bass, Grounded: Finding God in the World, a Spiritual Revolution (San Francisco: HarperOne, 2015).

74 **"We're at the end of the line":** Charles Clover, The End of the Line: How Overfishing Is Changing the World and What We Eat (London: Ebury, 2004).

Chapter 6: Hidden Treasures

77 **"Mouth-brooder":** E. Schwanck and K. Rana, "Male-Female Parental Roles in Sarotherodon Galilaeus (Pisces: Cichlidae)," Ethology 89 (1991): 229–243.

78 **During their reproductive period in the springtime:** S. Balshine-Earn and D. J. D. Earn, "An Evolutionary Model of Parental Care in St. Peter's Fish," Journal of Theoretical Biology 184 (1997): 423–431.

79 **Haflat samar:** Kenneth E. Bailey, Poet and Peasant: A Literary Approach to the Parables in Luke (Grand Rapids, MI: Wm. B. Eerdmans, 1976).

Chapter 7: It Is High Noon in the Desert

90 **"What is it you plan to do":** Mary Oliver, "The Summer Day," in *New and Selected Poems*, vol. 1 (Boston: Beacon Press, 1992).

Chapter 8: Farmsteads, Households, and Collaboration

94 **"We don't just live in the Empire":** Harvey Cox, *The Future of Faith* (New York: HarperCollins, 2009).

101 **Oikodespotes:** John Dominic Crossan, *The Power of Parable* (New York: HarperCollins, 2012). See also Crossan, *The Essential Jesus: Original Sayings and Earliest Images* (New York: HarperCollins, 1994).

Chapter 9: Farmworkers

106 **One of those was Norma Flores Lopez:** Norma Flores Lopez with Mariya Strauss, "Labor: I Was a Child Farmworker," AlterNet/Global Comment, accessed January 4, 2013, https://tinyurl.com/bdaenqz.

106 **Seven out of every ten farmworkers laboring in the US are foreign-born:** Steven Zahniser, *Farm Labor Markets in the United States and Mexico Pose Challenges for U.S. Agriculture*, EIB-201, U.S. Department of Agriculture, Economic Research Service, November 2018, https://tinyurl.com/yxl7r2rk.

107 **The death rate among farmworkers while on the job:** National Farmworker Ministry YAYA staff, *Farm Worker Issues: Health and Safety*, accessed July 21, 2020, https://tinyurl.com/y2qryovy.

110 **The word used for "poor" in this part of the Gospel of Luke:** Ellen F. Davis, *Biblical Prophecy: Perspectives on Christian Theology, Discipleship, and Ministry* (Louisville, KY: Westminster John Knox, 2014).

113 **Such crowded quarters are like Petri dishes:** Gary Nabhan, "Migrant Farmworkers, Native Ranchers in Border States Hit Hardest by COVID-19," Civil Eats, May 22, 2020, https://tinyurl.com/y6wuppmk.

Chapter 10: The Wonders of Weeds

120 **Its ability to mimic the appearance of wheat:** John Hutton Balfour, *Plants of the Bible* (London: T. Nelson and Sons, 1885).

123 **Would infuse the leaves of darnel grass:** James A. Duke, *Medicinal Plants of the Bible* (New York: Trado-Medic Books, 1983).

124 **This second fungus produces toxins:** L. P. Bush, H. H. Wilkinson, and C. L. Schardl, "Bioprotective Alkaloids of Grass-Fungal Endophyte Symbioses," *Plant Physiology* 114, no. 1 (1997): 1–7.

124 **They provide the darnel with a chemical defense:** K. Clay, "Fungal Endophytes of Grasses: A Defensive Mutualism between Plants and Fungi," *Ecology* 69 (1988): 10–16.

Chapter 11: The Wild Edges

130 **A mustard seed "tends to take over where it is not wanted":** John Dominic Crossan, *Jesus: A Revolutionary Biography* (New York: HarperCollins, 1994).

Chapter 12: Herders

139 **Between one hundred and two hundred million people:** Lisa Friedman, "U.N. Report Sees Climate among the Forces Ending Herders' Home on the Range," Environment and Energy, March 27, 2012, https://tinyurl.com/ydynnr72.

142 **Those who stand at the Radical Center:** William McDonald, "The Malpai Borderlands Group: Building the Radical Center," in Forging a West That Works: An Invitation to the Radical Center, ed. Barbara H. Johnson (Santa Fe, NM: Quivira Coalition, 2003), 3–10.

143 **Working hand in hand together on the land:** Dan Dagget, Beyond the Rangeland Conflict: Toward a West That Works (Layton, UT: Gibbs Smith Books, 2000). See also Susan Charnley, Thomas E. Sheridan, and Gary Paul Nabhan, Stitching the West Back Together: Conservation of Working Landscapes (Chicago: University of Chicago Press, 2014); and Courtney White, "A New Environmentalism," in Forging a West That Works, 53–80.

Chapter 13: Women Farmers

147 **In 2012, my home state of Arizona had the highest proportion of women farmers:** National Sustainable Agriculture Coalition, 2012 Census Drilldown: Women and Minority Farmers, accessed July 21, 2020, https://tinyurl.com/y5qf2xk4.

147 **We did not know that until the 2007 farm census:** Steven Manheimer, "Arizona Agriculture by the Numbers," Arizona Farm Bureau, March 1, 2009, https://tinyurl.com/yb6so6d8; G. P. Nabhan and J. Glennon, "The Changing Faces in Arizona's Food

Systems," green paper no. 1 (Tucson: Center for Regional Food Studies, 2016).

Chapter 14: Gleaners

154 **There were customs in place to regulate the practice of gleaning:** Lionel Casson, *Everyday Life in Ancient Egypt* (Baltimore: Johns Hopkins University Press, 1985).

157 **A number of Christian historians and theologians have linked this gleaning event:** Warren Carter, *Matthew and the Margins: A Sociopolitical and Religious Reading* (Sheffield, UK: Sheffield Academic, 2000).

158 **The Iskashitaa Refugee Network salvages tens of thousands of pounds per year:** Jonathan Bloom, *Harvesting Hope*, Iskashitaa Refugee Network, accessed July 21, 2020, https://tinyurl.com/y85w8m4h.

159 **The racial diversity as well as the very humanity of gleaners in my state:** G. P. Nabhan and J. Glennon, "The Changing Faces in Arizona's Food Systems," green paper no. 1 (Tucson: Center for Regional Food Studies, 2016).

160 **Another 23 percent was taken home by gleaners:** Anne Hoisington, Sue N. Butkus, Steven Garrett, and Kathy Beerman, "Field Gleaning as a Tool for Addressing Food Security at the Local Level: Case Study," *Journal of Nutrition Education* 33, no. 1 (2001): 43–48, https://tinyurl.com/yarzx7oa.

Chapter 15: Profiteers

166 **Reza Aslan unpacks the conditions:** Reza Aslan, *Zealot: The Life and Times of Jesus of Nazareth* (New York: Random House, 2013).

167 **This incident led directly to the crucifixion of Jesus:** Marcus J. Borg, Jesus: Uncovering the Life, Teachings, and Relevance of a Revolutionary (New York: HarperCollins, 2006).

167 **Jesus had gone to the heart of the problem of any corrupt religious institution:** John Dominic Crossan, Jesus: A Revolutionary Biography (New York: HarperCollins, 1994).

Last Suppers

169 **As Jay Parini reminds us:** Jay Parini, Jesus: The Human Face of God (Boston: New Harvest, 2013).

Chapter 16: Barren Figs

171 **The fruit of a fig tree could often be counted on:** Lytton John Musselman, Figs, Dates, Laurel, and Myrrh: Plants of the Bible and the Quran (Portland, OR: Timber Press, 2007).

174 **The taqsh themselves are sometimes thought of as "barren" figs:** F. F. Bruce, New Testament Documents: Are They Reliable? (Grand Rapids, MI: Wm. B. Eerdmans, 1943); W. M. Christie, Palestine Calling (London: Pickering and Inglis, n.d.).

174 **The rarer asexual process of parthenocarpy:** Mesut Baskaya and Julian C. Crane, "Comparative Histology of Naturally Parthenocarpic, Hormone-Induced Parthenocarpic, and Caprified Fig Syconia," Botanical Gazette 3, no. 4 (1950): 395–413.

Chapter 17: Communion

184 **Bread, wine, fish:** Cynthia Bourgeault, The Holy Trinity and the Law of Three (Boston: Shambhala, 2013).

184 **Jesus broke the rules of exclusion and purity:** Marcus J. Borg, *Jesus: Uncovering the Life, Teachings, and Relevance of a Revolutionary* (New York: HarperCollins, 2006).

FURTHER READINGS
ON PARABLES

Bailey, Kenneth E. *Jesus through Middle Eastern Eyes: Cultural Studies in the Gospels.* Downers Grove, IL: IVP Academic, 2008.

———. *Poet and Peasant: A Literary Approach to the Parables in Luke.* Grand Rapids, MI: Wm. B. Eerdmans, 1976.

Barnstone, Willis. *The Poems of Jesus Christ.* New York: W. W. Norton, 2012.

Borg, Marcus J. *Jesus: Uncovering the Life, Teachings, and Relevance of a Revolutionary.* New York: HarperCollins, 2006.

Capon, Robert Farrar. *Kingdom, Grace, and Judgment.* Grand Rapids, MI: Wm. B. Eerdmans, 2002.

Crossan, John Dominic. *The Essential Jesus: Original Sayings and Earliest Images.* New York: HarperCollins, 1994.

———. *The Power of Parable.* New York: HarperCollins, 2012.

Dodd, C. H. *Parable of the Kingdom.* London: Nisbet, 1935.

Donahue, John R. *The Gospel in Parable: Metaphor, Narrative, and Theology in the Synoptic Gospels.* Philadelphia: Fortress, 1988.

Levine, Amy-Jill. *Short Stories by Jesus.* New York: Harper and Row, 2014.

ABOUT THE AUTHOR

Brother Coyote, formerly known as Gary Paul Nabhan, is a professed member of the Order of Ecumenical Franciscans, an orchard keeper, seed saver, and agrarian activist. He has been a student in the Living School associated with the Center for Action and Contemplation and has practiced zazen at two Zen Centers in the Americas. He founded the Center for Regional Food Studies at the University of Arizona, where he holds the W. K. Kellogg Chair in Food and Water Security for the Borderlands. A former MacArthur Fellow whom the *Utne Reader* honored among its list of thought leaders making the world a better place, Nabhan is considered a pioneer in the heirloom seed saving, collaborative conservation, and local foods movements. In addition to writing or editing thirty books of poetry, essays, and stories, his scientific research and writings have appeared on National Public Radio and in *Nature*; the *Proceedings of the National Academy of Sciences*; *Plants, People, Planet*; and the *New York Times*. Following his Lebanese family traditions, he maintains an orchard of over a hundred rare fruit varieties and manages a fledgling nursery from his home in Patagonia, Arizona. With his wife, Laura Monti, he is also involved in community service, seasonal foraging, and habitat restoration in Comcáac (Seri) Indian villages on the desert shores of the Sea of Cortez in Mexico.